Green Thumbs Up

The Urban Gardener's Guide to Growing Your Raised-Bed Garden

Written: Dr. Russell Welch

Published by Remnant Warrior Publishing

11349 Skimmer Ct Jacksonville Florida 32225

"If you enjoy this book, would you be so kind as to help us remain in the top listing by leaving a positive Book review."

Contents

Preface

"The greatest fine art of the future will be the making of a comfortable living from a small piece of land."

- Abraham Lincoln

In an era where convenience often overshadows connection, the art of gardening invites us to slow down and engage with the simplicity and richness of life. "Green Thumbs Up: The Urban Gardener's Guide to Growing Your Raised-Bed Garden" is a heartfelt invitation to embark on a journey of cultivation and nourishment in your own backyard, or even on your balcony.

This book is crafted to guide you through the process of establishing a raised-bed garden, tailored specifically for those who have always believed that their limited space or urban setting bars them from the joys of gardening. Through clear, practical instructions and supportive insights, it demystifies the perceived complexities surrounding gardening. By turning small patches of potential into thriving gardens, this guide promises not only a bountiful harvest but also a renewed sense of connection with the environment.

The inspiration to write this book sprang from witnessing numerous individuals in urban settings express a yearning to grow their own food but retreat at the thought of overwhelming challenges. I recall a young couple from Orlando who shared their frustrations over their failed gardening attempts due to lack of space and knowledge. Their story, like many others, is filled with initial excitement followed by disappointment. It became evident that there was a deep need for a simple, yet comprehensive guide that could transform doubts into success.

In creating this book, I drew upon my diverse experiences in team building and strategic planning, as well as my profound commitment to spreading hope and grace as taught in the Gospel. My journey as a Pastor and leader has always been about nurturing growth, be it spiritual or in this case, agricultural.

I extend heartfelt gratitude towards my family, colleagues at Highway To Heaven Church, and the many gardening enthusiasts whose shared experiences have enriched this guide. Their stories and successes have been instrumental in shaping its content.

To you, the reader, thank you for choosing to invest your time with this book. Whether you're looking out over a sprawling yard or just a modest patio space, remember that every great journey begins with a single step—or in this case, a single seed.

This book is tailored for beginners but is also packed with insights that even seasoned gardeners might find enlightening. No prior experience is needed; just a willingness to learn and a desire to connect more deeply with the food you consume.

As you turn these pages, I invite you to embrace the journey of becoming an urban gardener. Let's cultivate not just gardens, but resilience and joy in spaces where both seem scarce. By the end, I hope you will not only enjoy the literal fruits of your labor but also the increased peace and satisfaction that comes from growing them. So here's to taking that first step—may it lead you to bountiful harvests and new beginnings in your own slice of nature.

CHAPTER 1

Small Spaces, Big Harvests

In the early hours of a crisp Saturday morning, Ellie found herself standing on her small balcony, a cup of steaming coffee in hand as she surveyed the limited space with a mixture of frustration and hope. The city sprawled around her apartment, a jungle of concrete and glass that offered little respite for someone yearning to feel soil under her fingers. She longed for a garden, a tiny oasis where she could cultivate life amidst the urban expanse.

Her thoughts wandered to her grandmother's garden, a lush canvas of colors and scents that had filled her childhood summers with joy. The tomatoes there had tasted like sunshine, utterly unlike the pallid varieties from the grocery store. Could she replicate even a fraction of that in such confined quarters? The idea seemed as distant as the rural tranquility of those days.

As she sipped her coffee, Ellie remembered reading about raised bed gardening. It was said to be ideal for situations like hers: compact spaces needing optimized soil and efficient water management. Maybe this could be her solution; perhaps she could coax life from wooden boxes perched above the city's hard surfaces.

She set her cup down and touched the cold railing, its metal chilling her fingertips. Below, people bustled along the sidewalks, wrapped up in their own morning routines. None glanced upwards towards her solitary figure framed against the skyline; none saw the battle between doubt and determination playing across her face.

Ellie pulled out an old notebook filled with sketches and notes on various apartment-friendly gardening techniques. Raised beds could improve soil quality far beyond what this concrete wilderness offered naturally. They would drain better than any pot could, reducing risks of root rot which had claimed more than one of her potted plants last season. And weeds—those relentless invaders—would be easier to manage in such an environment.

She imagined turning this barren balcony into a verdant patch where herbs scented the air and vegetables thrived within arm's reach. The challenge lay not just in constructing these raised beds but in maintaining them amidst unpredictable urban conditions—how much sun would be enough when skyscrapers cast long shadows?

Could this small space support life abundant enough to bring back not just flavors but memories? Would strawberries here ever taste as sweet as those plucked from sprawling fields under open skies?

Transform Your Tiny Space into a Thriving Garden

Gardening in small spaces can often seem challenging, but with the right approach, it becomes not only possible but also incredibly rewarding. Raised bed gardening is a perfect solution for urban dwellers with limited outdoor areas. This method allows you to create a productive garden in places as small as balconies or tiny backyards. It's about making the most of what you have and turning any small patch into a lush, vegetable-producing haven.

Raised beds offer **enhanced soil quality**, **improved drainage**, and **simpler weed management**, transforming gardening from a laborious task to an enjoyable activity. These beds can be filled with soil that is ideal for plant growth, which is often not the case with the native earth found in urban settings. By controlling the soil quality, you ensure that your plants get the nutrients they need without the competition from weeds.

For those new to gardening, constructing a planting box might sound daunting. However, it's simpler than it appears. The process involves choosing materials that are durable and safe for food crops, designing the box to fit your space, and ensuring proper drainage to keep plants healthy. This hands-on project not only provides a home for your plants but also adds a personal touch to your living space.

The sustainability of raised bed gardens is another significant advantage. These gardens require less water compared to traditional garden plots due to their containment, which reduces water waste. Additionally, by growing your own vegetables and herbs, you contribute to reducing the carbon footprint associated with transporting food items from farms to supermarkets.

This approach aligns perfectly with sustainable living practices and offers a practical introduction to organic gardening. By controlling what goes into your soil, you can avoid harmful chemicals and grow truly organic produce that is both healthy for your family and beneficial for the environment.

Throughout this book, we will explore various aspects of raised bed gardening—from selecting the right materials for your boxes to choosing plants that will thrive in your specific conditions. The aim is not just to guide you through setting up your garden but also to inspire confidence in growing and harvesting your own food.

By embarking on this journey of raised bed gardening, you embrace more than just a hobby; you adopt a lifestyle that values sustainability, self-reliance, and wellness. Whether you're looking to grow crisp lettuce or aromatic herbs, these small spaces can yield big harvests with a bit of planning and passion.

Remember, every large garden started as just a small seed—your gardening journey begins with a single step towards building your raised bed. Let's nurture that seed together into a flourishing garden that brings joy and nourishment into your life.

Understanding the Benefits of Raised Bed Gardening in Limited Spaces

Raised bed gardening is a beacon of hope for those who dream of cultivating their own garden but are constrained by the scarcity of space. This method involves the creation of elevated boxes filled with soil, allowing for plant cultivation in areas as small as balconies or small patches in urban settings. The essence of this technique is its ability to transform underutilized or unsuitable spaces into productive green havens.

Imagine your living space as a blank canvas, and raised beds as frames within which you can paint vibrant strokes of greens, reds, and yellows with vegetables and flowers. This not only maximizes the use of limited areas but also brings the joy and health benefits of gardening closer to home. The approach tailors gardening to fit into modern lifestyles where outdoor space is often a luxury.

One significant advantage of raised bed gardening is the control it offers over the growing environment. Traditional ground-level gardens are at the mercy of the soil beneath them, which might be poor in nutrients or contaminated. With raised beds, you can fill your planting box with tailor-made soil mixtures, ensuring your plants get all the nutrients they need without the risk of soil-borne diseases affecting your urban oasis.

Moreover, these elevated beds are easier on the back and knees. Gardening becomes more accessible since there's less bending and stooping involved. This feature is particularly beneficial for older adults or those with physical limitations who still wish to engage actively in gardening. By bringing the garden up to a more comfortable level, raised bed gardening invites more people to participate in this rewarding activity.

In summary, raised bed gardening turns small spaces into efficient and productive growing areas, making it an ideal solution for urban dwellers seeking to cultivate their own gardens.

Discover How Raised Beds Can Enhance Soil Quality, Drainage, and Weed Control

The introduction of raised beds into a garden setting fundamentally changes the dynamics of growing plants. Primarily, these structures offer improved soil quality — a vital factor for healthy plant growth. In raised beds, gardeners can customize soil blends tailored specifically to the needs of their plants, incorporating compost and other organic materials that enrich soil fertility and structure.

Consider this: if a gardener were a chef, then the soil in their raised beds would be akin to their signature dish — carefully prepared with just the right ingredients for perfection. This personalized approach ensures that plants thrive in an environment suited just for them, increasing yield and health.

Furthermore, drainage is significantly enhanced in a raised bed setup. Since these beds are elevated above ground level and often filled with loose, airy soil mixes, water logs less frequently than in traditional gardens. This prevents root rot and other moisture-related diseases which can devastate plant life.

The control over weeds is another compelling reason to adopt raised bed gardening. With clearer boundaries between walkways and growing areas, it's easier to manage unwanted vegetation. The height of raised beds also makes it more difficult for weeds to invade from nearby landscapes or beneath the surface.

In addition to these practical benefits, there's also an aesthetic appeal in using raised beds. They provide a neat and organized structure that can make even the smallest spaces look tidy and planned.

Could understanding these enhancements be what inspires you to start your own raised bed garden?

Learn the Basics of Constructing a Planting Box for Small Balconies or Backyards

Constructing a planting box for your balcony or backyard might seem daunting initially but breaking it down into manageable steps makes it achievable for everyone. First, determine the size based on your available space and what you intend to plant. Standard dimensions might include lengths ranging from 3 feet to 8 feet long while being about 2 feet wide.

The materials used can vary from wood (such as cedar or redwood) which naturally resists rotting, to more sustainable options like recycled plastic or composite materials that offer durability without harming the environment.

Visualize this: each plank or panel is like a building block in a game — where each piece you add helps form your very own miniature ecosystem right on your balcony or backyard corner.

Assembling these materials requires basic tools such as a saw for cutting wood (if necessary), screws or nails for joining parts together securely, and perhaps non-toxic sealants to protect wooden surfaces against weather elements. Adding drainage holes at the bottom and lining it with landscape fabric helps ensure excess water escapes without washing away valuable soil nutrients.

By following these steps carefully, anyone can create an effective planting box that caters precisely to their needs while fitting perfectly into limited outdoor spaces.

Raised bed gardening enhances urban living by converting small spaces into productive gardens; improves growth conditions through better soil quality; simplifies construction for personal adaptability; creating valuable opportunities for sustainable living right at home.

Raised bed gardening is not just a functional choice; it's a transformative one. For those of you with limited space, embarking on this journey means unlocking the potential of every square inch of your balcony or backyard. **The benefits are clear**: improved soil quality, better drainage, and simplified weed control enhance your gardening experience, making it more fruitful and less cumbersome.

Embrace the opportunity to grow your own food sustainably. Imagine the joy of harvesting fresh vegetables and herbs just steps from your kitchen. This method isn't just about utility but also about the pleasure and satisfaction that comes from tending to a garden. Each plant you nurture is a step towards a healthier lifestyle and a greener planet.

Constructing your own planting box might seem daunting at first, but it's an empowering first step. It's a testament to what can be achieved with a bit of wood, some soil, and seeds—even in the smallest of spaces. You'll learn that these physical boundaries don't define your gardening capabilities; rather, they set the stage for creativity and innovation.

As we move forward in this book, keep in mind that each chapter builds on these foundational concepts. You'll gain not only practical skills but also insights into making the most of what you have, wherever you are. Whether you're a seasoned gardener or a curious newcomer, there's something valuable here for everyone.

Let's cultivate not just gardens but also resilience and joy in our lives through gardening. Together, we'll explore how even the tiniest patch of earth can be turned into a thriving garden full of life and bounty. Get ready to put your green thumb to work and watch as your little space yields big harvests!

CHAPTER 2

Green Beginnings: You Don't Need a Farm

In the early hours of a brisk spring morning, Ella stood in the small patch of earth that lay behind her city apartment, a scant balcony really, but to her, it was a burgeoning canvas. The sun had not yet climbed above the horizon, but the sky hinted at a soft blue. Her hands felt the chill of dawn as she held a packet of seeds—basil and tomatoes—gifts from an old friend who insisted that "anyone can grow a garden."

She remembered his words clearly, spoken with such ease that they almost seemed trivial. "You don't need acres, Ella. Just a few pots and some love." Love she had in abundance; it was space and knowledge she doubted.

As she filled small pots with soil—a mix she had researched tirelessly online—the coarse texture rubbed against her fingers, grounding her thoughts. She thought about her grandmother's garden, sprawling and wild, each plant thriving under her experienced hand. Could she replicate even a fraction of that lushness here? The doubts hummed at the back of her mind like bees around a hive.

The city woke slowly around her. A car horn sounded in the distance; a dog barked twice then fell silent. Ella pressed seeds into the soil with gentle fingers. The simple act felt like both an ending and a beginning.

Her neighbor's window opened above her balcony, coffee aroma mingling with the fresh morning air. "What are you planting?" called down Mrs. Gilbert, an elderly woman with eyes like sharp sapphires.

"Just trying my hand at some herbs and tomatoes," Ella responded, unsure why admitting this made her feel vulnerable.

Mrs. Gilbert laughed softly—a comforting sound—and offered unsolicited yet welcomed advice about watering schedules and sunlight hours. It was advice born from experience; Ella soaked it up as eagerly as the parched earth soaking up the first water she sprinkled over her newly planted seeds.

Later that day, sitting by her window that overlooked this tiny green space, Ella found herself envisioning future meals flavored with her own herbs and vegetables. The fantasy was vivid: plucking basil leaves to scatter over homemade pizza or slicing fresh tomatoes for a salad.

But would these plants thrive? Could they defy urban concrete to reach their potential? She pondered over what success in this venture might look or taste like—would it be merely surviving or truly flourishing?

Isn't it curious how something as simple as planting seeds can stir hopes for growth not just in gardens but within ourselves too?

From Concrete Jungles to Green Marvels: Begin Anywhere, Grow Everywhere

Gardening is often perceived as an endeavor reserved for those with sprawling yards and ample time. However, this chapter challenges that notion head-on, illustrating that **you don't need a farm** or expert knowledge to cultivate a lush, productive garden. Whether you have a small balcony, a modest backyard, or just a few windowsills, the potential for growing your own plants is immense and achievable.

The first step in transforming your perception of gardening involves debunking the myth that it requires extensive land and expertise. Urban environments are teeming with untapped gardening opportunities that don't demand traditional spaces. Raised bed gardens, container gardens, and vertical planters are just a few examples of how you can make the most of limited space while ensuring good soil quality and drainage. These methods not only optimize small areas but also reduce the physical strain of gardening, making them ideal for beginners.

Embrace Simplicity: Your Green Thumb Awaits

Next, it's crucial to embrace simplicity in your gardening approach. Start by planning your garden layout—consider how much space you have and what you want to grow. Choosing plants suited to your climate and space limitations is key. For instance, herbs like basil and mint or vegetables like tomatoes and peppers thrive well in confined spaces and require minimal maintenance. Step-by-step instructions on setting up raised beds or containers can significantly ease the process, helping you focus on enjoying the growth journey rather than getting overwhelmed by complexities.

Cultivating Confidence: Small Steps to Big Green Gains

Starting small is an excellent strategy for novice gardeners. Begin with a few pots or a single raised bed. This manageable scope allows you to learn about plant care basics such as watering techniques, light requirements, and soil consistency without feeling overwhelmed. As you gain more confidence and experience from these initial successes (and failures), you can gradually expand your garden, experimenting with new plants and gardening methods.

Celebrate Every Leaf and Blossom

Celebrating progress is vital in keeping your gardening spirits high. Each sprout, leaf, or vegetable harvested should be seen as a triumph. Gardening is not just about the end produce; it's about the learning experiences along the way—the quiet joy of seeing your plants grow day by day and knowing they thrive under your care.

Garden Steps Simplified: *From Novice to Nature Nurturer*

Goal: Empower anyone to start their own garden regardless of space limitations or prior knowledge.

1. **Debunk Myths**: Challenge common misconceptions; introduce successful small-space gardening examples.
2. **Plan & Design**: Offer clear guidelines on creating functional garden layouts even in limited spaces.

3. **Select Plants & Soil Care**: Assist in choosing suitable plants; provide easy-to-follow soil preparation steps.
4. **Progress & Patience**: Encourage celebrating small wins; emphasize continuous learning.

With each step taking roughly one week to implement—starting from myth-busting to celebrating initial growths—you can see tangible results in just one month.

Successful gardening is not defined by scale but by adaptability and creativity in using available resources effectively. By breaking down barriers through practical guidance, this chapter inspires anyone with enthusiasm for gardening to start their journey toward sustainability and self-sufficiency today—proving that indeed, green beginnings can happen anywhere.

Debunking Gardening Myths

One common myth is that you need a vast expanse of land and a deep reservoir of knowledge to start gardening. However, this isn't true. Urban gardening has shown that even a small balcony or windowsill can become a lush garden. You don't need acres; a few pots or raised beds are sufficient to grow herbs, vegetables, and flowers.

In the heart of the city, think of each little green space as an oasis in a desert of concrete. Just as an oasis thrives in limited space, so can your urban garden. This analogy illustrates that with the right conditions and care, life finds a way to flourish, no matter the size of the plot.

It's also assumed that one must be armed with extensive horticultural knowledge before planting the first seed. Yet, many gardeners start with basic information and learn through experience. Resources are plentiful—from books to community workshops and online tutorials—that can guide you step by step.

Every plant started as a seed that needed nothing more than soil, water, and sunlight to begin its life journey. Similarly, every gardener starts with just a basic curiosity about growing something green. From there, gardens bloom with the addition of knowledge and experience gathered along the way.

Gardening is accessible to everyone, regardless of space or prior knowledge.

Simple Steps to Start Gardening

Embarking on your gardening adventure doesn't require complex strategies; it begins with simple steps that anyone can follow. Firstly, identify what you'd like to grow based on your interest and local climate—be it aromatic herbs, fresh vegetables, or beautiful flowers.

Understanding the basic needs of your chosen plants is crucial. Most plants need adequate sunlight, water, and good soil to thrive. Start by ensuring these needs can be met in your chosen gardening spot, whether on a sunny windowsill or an outdoor container garden.

Next comes the fun part: getting your hands dirty! Planting seeds or seedlings is a straightforward process that brings immense satisfaction. Watching those first sprouts appear is magical and reinforces the joy of gardening.

Gardening also teaches patience and perseverance as you nurture your plants through challenges like pests or unpredictable weather. These experiences build not just your gardening skills but also personal resilience.

Lastly, consider how you'll enjoy the fruits of your labor—whether incorporating fresh herbs into cooking or creating bouquets from your own flowers adds another layer of enjoyment to the gardening process.

Could starting simple be the key to unlocking your green thumb?

Choosing Plants and Maintaining Soil

Selecting the right plants for your garden is more about matching plant needs with environmental conditions than about choosing what looks best. For beginners, it's wise to start with low-maintenance plants known for their adaptability to various conditions.

Soil quality cannot be underestimated—it's fundamental for plant health. Beginners might consider using high-quality potting mix which provides a balanced medium for most container plants. Learning about soil amendments like compost can further enrich your garden's foundation, promoting robust plant growth.

Imagine if each plant's roots were tiny mouths eating up nutrients from the soil—this metaphor helps underscore how vital nutrient-rich soil is for healthy plants.

In maintaining soil health, regular assessments are helpful. Checking for proper moisture levels and ensuring good drainage will protect roots from rot and keep plants thriving.

Starting simple with careful plant selection and soil management sets you on the path to gardening success.

These foundational steps ensure anyone can embark on their gardening journey confidently and joyously, fostering a connection with nature regardless of their living situation. By understanding that elaborate plots or expert knowledge aren't prerequisites for successful gardening, more people may feel encouraged to start their own green spaces—an empowering realization that nurtures both gardens and gardeners alike.

Gardening is a journey that begins with a single step, not an expansive plot of land or a deep well of knowledge. It is heartening to realize that the creation of a lush, productive garden is within your reach, no matter the size of your balcony or backyard. This revelation opens doors to new possibilities and underscores a profound truth: **growth starts small**.

The myths surrounding gardening can often seem as tangled as an overgrown vine. Yet, we've seen how these can be unraveled with straightforward, practical advice. Starting your garden requires little more than commitment and some basic understanding, both of which are well within your grasp. Remember, every expert gardener was once a beginner too.

Selecting the right plants and maintaining healthy soil might initially seem daunting. However, these tasks become less intimidating when approached with patience and persistence. The wisdom found in *Philippians 4:13*, "I can do all things through Christ who strengthens me," serves as a powerful reminder that you are not alone in this endeavor. Your garden, much like faith, flourishes with attention and care.

Let each seed you plant be a testament to your resilience and faith in the unseen. Every sprout that breaks through the soil is proof of what faith and hard work can achieve. As you nurture your garden, let it also nurture you—reminding you of the simple joys and profound lessons embedded in nature.

Embrace this green beginning with enthusiasm and confidence. The rewards of gardening extend far beyond the harvest; they reach into the very roots of spiritual growth and personal fulfillment. Your journey starts here, one seed at a time.

CHAPTER 3

Overcoming the Fear of the First Seed

Ella knelt beside the row of wilting tomatoes, her fingers brushing against the dry soil, fragments of earth clinging to her skin like desperate whispers. The sun was a relentless overseer, casting a harsh light over her small backyard garden in the heart of suburban Lancaster. It was mid-July, and the air hung heavy with the scent of overheated plants and Ella's growing frustration.

She had read books, scoured online forums for advice, and even attended local gardening workshops. Yet, despite her efforts, her garden seemed to mock her with its continuous array of problems: pests feasting on leaves as though they were at a banquet, mysterious fungal infections spreading like gossip, and now this—heatwaves that threatened to suck the life out of everything green.

As she adjusted her straw hat to shield her eyes better from the glare, a gentle breeze stirred around her—a temporary reprieve. Her thoughts drifted to last winter when planning this garden had been a source of bright dreams amid short days. She had imagined lush vines heavy with tomatoes and cucumbers; rows of carrots and beets pushing up from the earth like buried treasure. Those images had felt so real then but felt like distant fantasies now.

Nearby, the sound of children laughing drifted over from the neighbor's yard where a sprinkler cast arcs of water into the air. The simplicity in their joy struck a chord in Ella. Gardening was supposed to be her escape, her simple joy; why then did it feel like she was fighting a losing battle?

Determined not to let another day end in defeat, she stood up and fetched her watering can and notebook filled with gardening tips gathered over months. Flipping through pages smeared with soil fingerprints, she paused at a section about resilience in gardening—about understanding plant needs deeply and responding proactively rather than reactively.

The sun began its descent towards the horizon as she methodically watered each plant while whispering encouragements as if they could hear and understand her longing for them to thrive. The cool water seemed almost soothing after the heat of day; perhaps not all was lost yet.

As dusk painted shadows across her backyard sanctuary, Ella sat back on her heels amidst rows that bore both promise and disappointment so closely intertwined. Could it be that each failure taught something vital about patience, resilience? Could these struggling plants teach Ella not just about gardening but about nurturing herself?

Would tomorrow bring another lesson hidden beneath dew-covered leaves?

Will Your Garden Grow? Conquering the Fear of Planting the First Seed

Starting a garden can feel like stepping into unknown territory. Many potential gardeners hesitate at the threshold, daunted by "what-ifs" and fear of failure. This hesitancy is natural, but it's also surmountable. Gardening, much like life, is a journey of learning and growth. Every seasoned gardener started with a single seed and a bundle of uncertainties. The key to success lies not in avoiding failure but in embracing it as a stepping stone to knowledge and resilience.

The first step to overcoming gardening apprehension is **understanding common issues** that may arise and how to handle them. Knowledge is power, particularly when it comes to plant care. By learning about the most frequent challenges—be it pests, diseases, or nutrient deficiencies—you equip yourself with the tools needed to respond effectively. This doesn't just solve problems; it prevents the fear associated with potential failures from ever taking root.

Building confidence is next on your path to a flourishing garden. Confidence grows from experience and knowledge. Each chapter of your gardening story will add layers to your understanding, from choosing the right soil to knowing when to water and how much sunlight your plants need. Embrace each small victory—a sprouting seed, a thriving plant—and let these successes bolster your gardening self-esteem.

Furthermore, cultivating resilience is crucial. Resilience in gardening means accepting that some plants might not thrive despite your best efforts. This isn't failure; it's a part of the natural cycle and an opportunity for learning. Each setback teaches you more about your garden's specific needs and how to better care for your plants moving forward.

Remember that every gardener has faced setbacks. What separates those who succeed from those who give up is not luck but perseverance and the willingness to learn from mistakes. Approach gardening with curiosity and patience—qualities that are invaluable not just in gardening but in all walks of life.

To foster this resilience, start small. Choose easy-to-grow plants that are known for their hardiness. Herbs like basil and mint or vegetables like lettuce and radishes can be great starters that provide quick gratification and boost your confidence early on.

Lastly, always seek advice when needed and share your experiences with others. Gardening communities—both online and in local neighborhoods—are fantastic resources for support and knowledge sharing. Engaging with fellow gardeners can provide moral support, practical advice, and may even lead to lasting friendships rooted in shared interests.

By approaching gardening with an informed mind, a resilient spirit, and a supportive community, you'll find that planting that first seed isn't as daunting as it seems. It's the first step in a rewarding journey of growth, learning, and connection with nature.

Learning Strategies for Managing Common Gardening Issues and Failures

Gardening is not just about planting seeds and waiting for them to grow; it's about managing the many small challenges that arise along the way. One of the most common issues gardeners face is pest infestation. Simple preventive measures, such as regular inspection of plants and proper spacing to ensure good air circulation, can significantly reduce the risk of pests.

Imagine your garden as a small community where each plant relies on others for survival, much like a team in sports where every player has a role that supports the whole. Just as a coach might adjust strategies to protect players from injuries, gardeners must also adapt their approaches to safeguard their plants.

Another frequent concern is over or under-watering, which can lead to weak plants susceptible to diseases. A basic rule of thumb is to water deeply but infrequently, encouraging roots to grow deeper into the soil. This method helps plants become more resilient and better equipped to access nutrients they need to thrive.

Soil quality also plays a crucial role in gardening success. Enriching your soil with compost not only improves its structure but also increases its nutrient content. Regularly adding organic matter can transform even the poorest soil into a rich, fertile foundation that supports robust plant growth.

Effective gardening is about proactive problem-solving and regular care. By understanding and anticipating common issues, you can prevent many problems before they even start.

Strategies for resilience include preventive pest control, careful watering practices, and continuous soil improvement.

Building Confidence Through Practical Gardening Advice

Confidence in gardening grows from accumulated small successes and learning from everyday challenges. For instance, choosing the right plants for your climate zone is crucial. Local extension services and garden centers can provide valuable guidance on which plants are best suited to your area's conditions.

Have you ever noticed how a single positive experience, like successfully growing your first tomato plant, can make you feel more capable and motivated? It's similar in gardening; early successes with hardy, low-maintenance plants can boost your confidence dramatically.

Incorporating mulch into your garden not only conserves moisture and suppresses weeds but also teaches patience and foresight—laying mulch today means less weeding tomorrow. This kind of forward-thinking is essential in building both a beautiful garden and a confident gardener.

Practical advice often includes keeping a garden journal. Recording planting dates, weather patterns, and when pests appear helps you develop an intuitive understanding of your garden's rhythms and needs. Over time, this knowledge builds a more instinctive approach to gardening.

Success in gardening doesn't come from never failing but from becoming adept at responding constructively when things don't go as planned. Embrace each setback as an opportunity to learn and refine your techniques.

Could recognizing these small victories be the key to growing not just plants but your confidence as well?

Cultivating Resilience by Understanding Plant Care Basics

Resilience in gardening starts with understanding the basics of plant care. Knowing how much light different plants need allows you to position them optimally within your space. Most vegetables, for example, require at least six hours of direct sunlight per day to thrive.

Watering seems simple but understanding the needs of various plants can be complex. Some prefer dry conditions while others thrive in moist soil. Learning when and how much to water is fundamental in preventing stress on plants which can lead to disease or poor growth.

Feeding your plants appropriately is also critical. A general-purpose fertilizer may suffice at times, but tailored feeding that meets specific nutritional needs of different plants will yield better results. This targeted approach prevents nutrient excesses or deficiencies that can stress plants unnecessarily.

Each plant has its unique set of requirements and tolerances; think of it as knowing individual friends' preferences for comfort and care—some like high energy environments while others prefer calm and minimalistic spaces.

By investing time in learning these basics, you build a foundation not just for your garden's success but for personal resilience against setbacks. As each issue arises—be it pests, disease or poor weather—you're better equipped to handle it confidently.

Understanding plant care basics builds resilience in both gardens and gardeners; this knowledge empowers us to handle setbacks effectively and ensures our efforts lead to fruitful harvests.

Embarking on the journey of gardening can initially seem daunting. Fear of the unknown and potential failure often holds many back from planting that first seed. However, remember that **every**

gardener starts with a single step—a step of faith and commitment to learning and growth.

In managing common gardening issues, you've learned that setbacks are not the end but opportunities for learning and improvement. Just as in life, resilience is built through overcoming challenges. Scripture reminds us, "We are afflicted in every way, but not crushed; perplexed, but not driven to despair" (2 Corinthians 4:8). This resilience is what transforms a budding gardener into a seasoned green thumb.

Building confidence in your abilities doesn't come from never failing but from realizing that you have the capacity to recover and learn from each experience. The practical advice shared here aims to equip you with knowledge and tools to handle various situations confidently. It's about nurturing your skills as much as you nurture your plants.

Understanding the basics of plant care is akin to understanding how to feed and care for one's body or soul. It's fundamental. When you know what your plants need, you can provide for them better, leading to a healthier garden. This knowledge reduces anxiety and increases enjoyment in your gardening practice.

As we move forward, hold onto these insights and approach your garden with optimism and patience. Let each seed you plant be a testimony to your faith in new beginnings and growth. Remember, each plant that thrives under your care can be a reflection of the grace and love inherent in nurturing life.

Embrace this journey with an open heart and mind, ready to learn and adapt. Your garden is more than just a space for plants to grow—

it's a place where you can sow seeds of hope and harvest not just vegetables or flowers, but joy, peace, and satisfaction. Be encouraged by each small success and use it as a stepping stone towards greater achievements in your gardening adventures.

You are capable, *and with time,* will see the fruits of your labor bloom abundantly.

CHAPTER 4

Simplifying the Secrets of the Soil

In the quiet town of Cedar Grove, nestled between rolling hills and sprawling fields, Anna spent her mornings tending to her garden. It was a modest plot, framed by aged oak planks that had turned a soft gray from the sun's relentless kiss. The soil within these raised beds was dark and rich, a stark contrast to the hard clay that lay just beyond.

Anna's hands moved with a practiced ease, her fingers tracing the delicate stems of young tomato plants, encouraging them upright. The morning air was crisp, carrying the scent of earth and dew. Birds chirped from the branches overhead, their songs weaving through the rustle of leaves like threads in nature's own tapestry.

Her thoughts drifted to last season's struggle with pests—the relentless aphids and the cunning rabbits that had found their way into her neighbor's garden. But here, in her raised beds, such challenges seemed distant memories. The elevated soil deterred many critters, and those persistent enough to climb found themselves quickly discouraged by barriers easily installed at this manageable height.

As she adjusted a vine laden with nascent cucumbers, Anna recalled how overwhelming gardening once felt. Advice from every corner—organic this, hydroponic that—it had all muddied her enthusiasm like too much rain spoiling tender seedlings. Yet here among her raised beds, clarity found its way to her like sunlight breaking through after a storm. This approach simplified everything: better soil control meant healthier plants; fewer pests meant less despair.

The chime of the wind bell pulled Anna from her reverie as a gentle breeze passed through the garden. She straightened up and wiped her brow with the back of her hand, leaving a smudge of soil on her forehead unnoticed. A robin alighted near the birdbath, tilting its head as if curious about human concerns.

Anna smiled at it momentarily before casting an evaluative glance over her garden—her sanctuary where simplicity met productivity in beautiful symbiosis. Could this focused method not only nurture plants but also restore peace to overwhelmed souls?

Unearth the Simplicity: Mastering Gardening Basics with Raised Beds

Raised bed gardening is a method that both novices and seasoned gardeners can embrace with success. It distills complex horticultural concepts into manageable, bite-sized pieces. This approach not only makes gardening more accessible but also enhances the enjoyment and satisfaction derived from growing your own food or flowers. By focusing on raised beds, individuals can bypass common barriers to entry in traditional gardening, such as poor soil conditions and overwhelming maintenance requirements.

The design of raised beds inherently improves **soil quality**. With the ability to fill your beds with a customized soil mix, you can create an ideal environment for your plants to thrive. This control over soil composition is pivotal because healthy soil fosters healthy plants which are more capable of resisting pests and diseases.

Moreover, the elevation of raised beds ensures better drainage, which is crucial for preventing root rot and other moisture-related issues.

Pest control is another significant advantage provided by raised bed gardening. The elevated structure often deters many ground pests, and additional protective measures like hardware cloth can be integrated during the setup to prevent burrowing intruders. This means less reliance on chemical pesticides, aligning with environmentally friendly practices and promoting a healthier ecosystem in your garden.

For many beginners, the vast array of gardening philosophies can be daunting. Raised bed gardening simplifies this by providing a clear framework within which to operate. It encourages focusing on essential practices like crop rotation, proper spacing, and simplified weed management. These practices are easier to implement in a contained space, making garden management less overwhelming and more enjoyable.

The chapter also emphasizes the importance of viewing gardening as a nurturing process akin to fostering faith. Just as our spiritual journeys benefit from structured guidance through scriptures, our gardens thrive under the disciplined care facilitated by raised beds. This analogy resonates deeply within the faith community, underscoring how structured environments can lead to growth and prosperity—both spiritually and agriculturally.

Furthermore, embracing raised bed gardening can be seen as an act of stewardship towards God's creation, encouraging sustainable practices that care for the earth responsibly. It's about taking small steps towards making a big impact, reflecting principles of grace and mindfulness in every action we take in our garden spaces.

By turning to raised bed gardening, beginners are equipped with the tools they need to start their gardening journey on the right foot. They learn that successful gardening doesn't have to be complex or filled with esoteric knowledge but rather approached with patience, care, and a willingness to learn—values deeply rooted in faith-based teachings.

In essence, this chapter provides not just practical advice but also spiritual encouragement for embarking on a gardening adventure that nourishes both the body and soul. It invites readers to simplify their approach and focus on what truly matters: creating a thriving garden that reflects both their dedication and their love for nature.

Grasp How Raised Bed Gardening Clarifies Gardening Principles

Raised bed gardening serves as a simplifier in the often complex world of horticulture. For beginners, understanding the basic principles of gardening can be daunting. Raised beds, however, offer a straightforward framework within which these principles are more manageable. By confining the gardening space to a clearly defined area, raised beds help in focusing on essential aspects like soil management and planting techniques without the distractions of a larger garden.

Imagine a classroom where each student has their own desk with all necessary supplies within arm's reach, compared to a large auditorium where everyone shares resources. This is akin to the difference between raised bed gardening and traditional row gardening. In raised beds, everything the plant needs is readily available and tailored to its specific requirements, making it easier for the gardener to monitor and control growth conditions.

The structure of raised beds also means that soil can be optimized specifically for gardening. Unlike native garden soil, which may be compacted or mixed with undesirable materials, soil in raised beds can be blended to suit specific plant needs. This customization reduces confusion for new gardeners about why their plants may not be thriving and helps them learn how to adjust soil compositions effectively.

Another significant advantage is the reduced need for heavy tools and equipment. In traditional gardening, tilling and aerating large plots demand considerable physical effort and tool use. Raised beds typically require smaller, more manageable tools that are perfect for beginners who might feel overwhelmed by more heavy-duty equipment.

Raised bed gardening distills complex gardening principles into more understandable segments by providing a controlled environment that focuses on fundamental soil and plant interactions.

Explore the Advantages of Improved Pest Control and Soil Quality in Raised Beds

One of the standout advantages of raised bed gardening is enhanced pest control. Pests are less likely to reach plants in raised beds compared to those planted directly in the ground. The elevation creates a physical barrier that many crawling insects find difficult to navigate. Additionally, it's easier to attach netting or row covers to raised beds to protect against flying pests.

This improved pest management is not just about barriers; it also involves proactive strategies such as crop rotation and interplanting which are simpler to execute in a contained space. These methods disrupt pest life cycles and reduce infestations without resorting to harsh chemicals, making your garden safer and more sustainable.

When considering soil quality, raised beds provide an unparalleled advantage. Gardeners can tailor the soil mixture to suit specific plant needs, incorporating the perfect balance of nutrients and proper drainage capabilities. This ability ensures that plants have optimal growing conditions from the start, significantly boosting their health and yield.

Moreover, since gardeners have control over what goes into their raised bed soil, they can avoid problems associated with native soils such as clay or sand imbalances or contamination from previous chemical use. The result is not only healthier plants but also a safer environment for both the gardener and local wildlife.

Could understanding these benefits lead you to consider how much more effective your gardening efforts might be with raised beds?

Remove Overwhelm from Varied Gardening Philosophies with a Focused Approach

The world of gardening is rich with diverse philosophies that range from biodynamic agriculture to square foot gardening. For someone new to this world, sifting through these varied approaches can be overwhelming. Raised bed gardening offers a focused methodology that simplifies learning by concentrating on basic yet essential practices.

Consider this: if you were learning how to cook, you wouldn't start with advanced molecular gastronomy. Instead, you would focus on mastering fundamental recipes first. Similarly, raised bed gardening teaches foundational horticultural skills before moving on to more complex techniques.

This approach allows beginners to build confidence as they see tangible results quickly without getting bogged down by the intricacies of more comprehensive methods that require deeper knowledge and experience. Each success in their raised bed provides empirical evidence of what works best in their specific environment—knowledge that can later be applied elsewhere in broader gardening endeavors.

By reducing complexity, raised bed gardening helps new gardeners focus on mastering core practices such as proper planting times, spacing requirements, and basic pest management strategies—all crucial elements that form the backbone of successful horticulture.

By clarifying fundamental concepts through focused practices like those found in raised bed gardening, this method effectively removes confusion caused by conflicting gardening philosophies while setting up individuals for long-term success across various types of horticultural endeavors.

Raised bed gardening stands as a beacon of simplicity and clarity in the often complex world of horticulture. It strips back the unnecessary and focuses on what truly matters: **soil quality, pest control, and straightforward gardening principles**. For beginners, this method offers a gentle introduction to gardening without the overwhelm that often accompanies traditional methods.

Embrace Simplicity

By embracing raised bed gardening, you are choosing a path that is not only manageable but also deeply rewarding. The improved soil quality in raised beds ensures that your plants have the best environment to thrive. This is crucial because, as we learn from scriptures like Luke 8:15, "But the seed on good soil stands for those with a noble and good heart, who hear the word, retain it, and by persevering produce a crop." Just as good soil nurtures seeds into a fruitful harvest, so too does a well-maintained raised bed garden promise a bountiful return.

Harness Control

Moreover, better pest control in raised beds means less frustration and more joy in your gardening journey. It's about stewardship—taking control of your small patch of earth and nurturing it carefully. This mirrors the Biblical principle of stewardship found in Genesis 2:15, "The Lord God took the man and put him in the Garden of Eden to work it and take care of it." In managing our gardens effectively, we honor this call to care for creation.

Focus Your Efforts

Focusing on raised bed gardening also helps you filter out the noise from various gardening philosophies that can be distracting or contradictory. It allows you to concentrate on proven strategies that yield results, giving you peace and satisfaction in your gardening efforts. This approach is akin to Matthew 6:33, "But seek first his kingdom and his righteousness, and all these things will be given to you as well." In focusing on solid principles, all other benefits will naturally follow.

Raised bed gardening is not just about cultivating plants; it's about cultivating faith, hope, and love through the simple acts of planting, tending, and harvesting. It's a physical act of faith—a belief in the unseen growth beneath the soil—and an exercise in hope for future abundance. Most importantly, it's an act of love for the piece of Earth under your care.

Let this method simplify your gardening journey. Let it bring clarity amid confusion and success where there has been struggle. As you move forward with your raised beds, remember that each seed planted is not just potential for new life but also an affirmation of faith in tomorrow's harvests.

CHAPTER 5

The Essentials of Abundance: Skills for Every Season

In the early hours of a crisp spring morning, Thomas walked through his small but burgeoning garden, his boots sinking slightly into the moist earth with each step. The sun, a shy yet bold artist, painted the sky in hues of orange and pink. Around him, the world was waking; birds chirped their morning songs and a gentle breeze rustled through the young leaves of newly planted trees.

Thomas had always found solace in his garden. It was here among the rows of tender green shoots that he could escape the clutter of his mind and the weight of past decisions. This year, however, his garden bore not just the promise of escape but an urgent need for sustenance. With recent layoffs at the factory where he worked looming over him like dark clouds before a storm, each plant growing in this patch of earth was more than just potential beauty; it was potential sustenance.

He paused beside a row where he planned to implement crop rotation—a technique he hoped would yield better harvests than previous years. His fingers traced the edge of an empty plot where last year's tomatoes had drained the soil of its nutrients. Now, legumes would take their place, hopefully enriching the earth as they grew. His mind flickered to memories of his grandfather explaining how each plant takes and gives back differently to the soil—a dance of give and take that if mastered, promised abundance.

Moving to another section filled with young seedlings under protective coverings, Thomas thought about companion planting—the strategic placement he had studied late into several nights. He planted marigolds among tomatoes to ward off pests naturally; an old farmer's trick mingled with hope that fewer chemicals would mean healthier food for his table.

As he adjusted a sprinkler to ensure even watering—a tactic critical in avoiding both drought and root rot—he couldn't help but overhear his neighbor's loud complaints about aphids ruining her roses. Pest management was another battlefield for Thomas; one where victory meant survival in more ways than one this year.

Each step around his garden was both meditation and strategy session as Thomas mulled over these techniques that were now more necessity than hobby. He felt a twinge of fear at what failure might mean this season but pushed it aside with a deep breath that smelled like dew and growing things.

As Thomas looked out across rows soon to burst forth with life under his careful watch, one wonders: Can careful planning and ancient wisdom truly turn nature's favor?

Unlocking the Secrets to a Flourishing Urban Garden

Gardening in the urban environment presents unique challenges and rewards, making the mastery of certain foundational skills not just useful, but essential. In a landscape where space is at a premium and every inch of soil counts, understanding the intricacies of crop rotation, companion planting, and seed starting can transform a modest patch into a thriving garden. This chapter delves deep into these critical techniques, offering practical guidance to ensure your garden is not only surviving but flourishing.

Crop rotation is more than just a traditional farming practice; it's a strategic approach that keeps the soil healthy and nutrient-rich. By rotating crops according to their nutrient requirements and pest relationships, you can naturally enrich your soil without relying heavily on chemical fertilizers. This practice reduces the buildup of pests and diseases, ensuring that your garden remains productive season after season.

Companion planting, meanwhile, is an art form in itself. It involves placing plants together that can benefit each other either by enhancing growth, repelling pests, or attracting beneficial insects. This method not only maximizes space but also increases the biodiversity of your garden, which is crucial for ecological balance in an urban setting.

Starting from seeds may seem daunting at first, but it offers numerous benefits including cost savings and access to a wider variety of plant species. Understanding the right conditions for seed germination and the best practices for nurturing young seedlings are skills that will serve any gardener well, particularly when working within the confines of limited urban space.

Step-by-Step Guide to Urban Garden Mastery

Step 1: Mastering the Techniques

The journey to a bountiful garden begins with **mastering fundamental techniques**. Crop rotation should be planned based on family groups of plants to avoid depleting the soil of specific nutrients and to thwart common pests. For example, avoid planting tomatoes in the same bed immediately after peppers or potatoes since they are all susceptible to similar diseases.

Companion planting can be as simple as placing marigolds among your vegetables to deter pests with their strong scent or planting basil near tomatoes to enhance flavor and growth. These pairings help create a natural barrier against pests while promoting a healthier crop.

When starting seeds, use clean containers and quality soil designed for seed starting. Ensure seeds have adequate moisture and light—two critical factors that can make or break seedling vitality.

Step 2: Preparing the Soil

Healthy soil is the backbone of any successful garden. Begin by clearing your beds of weeds and debris from previous plantings—a clean slate encourages healthy new growth. Enhancing your soil with compost not only improves its structure but also increases its nutrient content. Testing your soil's pH can help you understand its condition better; most vegetables thrive in slightly acidic to neutral pH levels.

Step 3: Watering and Pest Management

Effective watering tactics do more than just keep plants alive; they encourage deep root growth which enhances plant health and drought resistance. Water deeply but infrequently to promote root health, adjusting your schedule based on rainfall and temperature.

For pest management, favor organic methods like introducing natural predators (ladybugs for aphid control) or using barriers (row covers) to protect plants from insects. These strategies reduce reliance on chemical pesticides, keeping your garden safe and sustainable.

Step 4: Ensuring a Bountiful Harvest

Preparation for each season involves revisiting these steps—rotating crops, preparing soil anew, adjusting watering schedules based on seasonal needs, and being vigilant about pest control methods. Regular monitoring helps catch issues early before they become problematic.

By following these structured steps consistently throughout each gardening season, you ensure not only survival but prosperity in your urban garden space. The joy of harvesting your own produce is unmatched—providing fresh ingredients right from your doorstep while fostering a deeper connection with nature even in an urban landscape.

Mastering Crop Rotation, Companion Planting, and Seed Starting

Crop rotation is an age-old practice that remains a cornerstone of sustainable agriculture. By rotating the types of crops grown in specific areas of your garden, you can naturally reduce pest populations and manage soil fertility. This technique involves changing the type of crop in a particular area with each growing season, thereby preventing pests and diseases that specialize in certain plants from establishing a stronghold.

Imagine your garden as a stage where different plays are performed each season. Just as audiences might tire of seeing the same play repeatedly, the soil can become exhausted by hosting the same crop over and over. Crop rotation keeps the soil healthy and full of nutrients, much like a varied theater program keeps its audience engaged and returning.

Companion planting is another key strategy that enhances the growth and health of your garden's plants. This method involves placing plants together that can benefit each other by deterring pests, enhancing growth, or even improving flavor. For instance, marigolds release a natural chemical from their roots that repels nematodes and other pests, which can protect neighboring vegetables.

Seed starting indoors allows gardeners to get a head start on the growing season. This control over the growing environment can lead to stronger, more resilient plants. By starting seeds indoors, you ensure your plants are strong enough to withstand outdoor conditions once transplanted. It's akin to nurturing young athletes in a training camp before they compete in major events – they perform better because they've had a strong preparatory phase.

Bold mastering these foundational techniques ensures a robust start for your garden each year.

Delving into Soil Preparation, Watering Tactics, and Pest Management

Soil preparation is fundamental for a thriving garden. It involves amending the soil with organic matter, ensuring proper pH levels, and establishing good drainage. These steps create an ideal environment for roots to grow deep and strong. A well-prepared soil is like a well-made bed: it's inviting and supports restful growth.

Watering tactics are critical in urban gardening due to variable natural water availability. Efficient watering practices such as drip irrigation or soaker hoses minimize waste and ensure water goes exactly where it's needed — at the roots. Overhead watering can lead to fungal diseases and should be avoided especially during evening times when moisture can linger on leaves overnight.

Pest management in an urban garden requires vigilance and strategic action. Instead of relying heavily on chemical pesticides, integrated pest management (IPM) approaches such as using beneficial insects or mechanical barriers can be effective. For example, installing netting or floating row covers can prevent pests from reaching plants while allowing light and air to penetrate.

Imagine you're setting up defenses in a medieval castle; similarly, protecting your garden from pests involves layers of strategy beyond just brute force.

Ensuring Preparedness for a Flourishing Season

Preparation is key to taking full advantage of your gardening efforts. It includes understanding what your plants need throughout their lifecycle—from sprouting seeds to harvesting fruits or vegetables. Being proactive about potential challenges like sudden weather changes or pest invasions can make all the difference.

A flourishing garden is much like preparing for a festive event—it requires attention to detail, anticipation of needs, and readiness for unexpected situations. For example, just as you might prepare extra food or seating for guests at a party, in gardening you might prepare extra support stakes or have organic pest repellents at hand should an unexpected infestation occur.

By staying informed about weather patterns and adjusting care accordingly—like providing shade during unexpected heatwaves or increasing water during dry spells—you safeguard your plant's health against climatic adversities.

Ensuring preparedness encompasses mastering crop rotation techniques for soil health; utilizing companion planting for natural benefits; starting seeds under controlled conditions; preparing soil meticulously; adopting efficient watering practices; and implementing thoughtful pest management strategies—all these contribute significantly towards achieving abundant harvests each season

As we wrap up our exploration of the essential skills for every urban gardener, let's remember that mastering techniques like **crop rotation**, **companion planting**, and **seed starting** is not just about following steps—it's about nurturing a relationship with the earth and reaping the rewards through abundant harvests. Each of these practices holds a promise similar to the biblical principle of sowing and reaping: what we invest in our soil, so shall we gather in time.

Preparing your soil adequately, understanding effective watering tactics, and managing pests wisely are foundational to creating a thriving garden. Just as we are called to care for our bodies as temples, so too must we care for our gardens as sacred spaces where life burgeons. By investing time and effort into these practices, you are setting the stage for a season filled with growth and bounty.

Lastly, ensure you are always prepared for each stage of your gardening journey. Like the wise builders from Matthew 7:24-27 who built their house on the rock, laying a strong foundation in your garden ensures stability and yield no matter what season lies ahead. Whether you face drought or deluge, your preparedness will reflect in the health and abundance of your harvest.

Remember, gardening is more than just a hobby; it is a stewardship. It teaches us patience, diligence, and care. It connects us to the cycle of life and reminds us of our role in nurturing and sustaining the environment. Embrace these lessons with joy and anticipation for what each new day in your garden may bring.

May your gardening efforts be fruitful and your harvests plentiful. Let each seed planted be a testament to faith in tomorrow's growth and abundance.

CHAPTER 6

The Green Pride: Nurturing Your Garden Oasis

Martha's hands delved into the cool, yielding earth, her fingers threading through the soil like roots seeking nourishment. It was early morning in her small garden, a patch of reclaimed land on the edge of an urban sprawl that had once known nothing softer than concrete. The sun, a gentle yet persistent presence, cast long, slender shadows that danced between the rows of burgeoning greenery.

She paused, a tomato seedling held tenderly between her fingers, and let her gaze drift across the fruits of her labor. Each plant told a story—a narrative of struggle and triumph so intricately tied to her own. Only months ago, this garden was but an idea; a fleeting escape from the clutches of processed foods and untraceable origins that filled supermarket shelves.

The scent of fresh earth mingled with the distant aroma of rain promised by dark clouds gathering at the horizon. Martha remembered her grandmother's garden—the way it seemed to pulse with life under her wise hands, each vegetable and herb a testament to patience and care. That memory spurred a longing in Martha not just for homegrown food but for connection; connection to her past and to the very essence of life that thrives beneath our feet.

She planted the seedling with care, her thoughts meandering through possibilities. Could this small act of nurturing life instill in others what it had in her? A respect for nature's cycles and an understanding that what we eat threads us inevitably back to the earth from which we all spring?

A neighbor passed by on the sidewalk beyond her fence, offering a wave and an admiring glance at the verdant oasis Martha cultivated from decay. "Looking good," he called out.

"Thank you!" she replied, wiping sweat from her brow with the back of her hand. His words were simple yet they swelled within Martha like music; they were recognition—not just of her garden's flourishing but also of her role in reshaping what was once mere space into sustenance.

As she stood there amid rows lined with promise—a future harvest that would feed both body and soul—Martha pondered whether this plot could serve as more than just a source of meals. Could it become a sanctuary for those like herself who yearned for something tangible among the intangible chaos of modern life?

What more might grow from these seeds she planted today?

Cultivating More Than Just Plants

Gardening, at its core, is about more than just tending to plants—it's a profound exercise in stewardship and personal growth. In the bustling pace of urban life, creating a garden oasis in your raised-bed can transform not just your living space but also your spiritual and emotional well-being. Through the nurturing of seeds and soil, you foster a unique connection with the creation, celebrating the empowerment that comes from growing your own food.

When you plunge your hands into the soil, you are doing more than planting; you are laying down roots in a process that mirrors personal and communal growth. This chapter guides you through understanding how gardening can be a source of deep joy and satisfaction. It's about harvesting more than vegetables or flowers; it's about cultivating resilience and pride in what you can achieve with patience and faith.

From Seed to Supper: The Journey of Empowerment

Growing your own produce is an empowering experience. Each seed planted is a testament to hope and expectation. As these seeds sprout and grow, so does an understanding of our role as caretakers of God's creation. This responsibility brings with it a sense of empowerment—knowing that the sustenance provided by the earth can be cultivated by our own hands reaffirms our capability and strength.

This empowerment extends beyond the garden. It influences how we view our food, making mealtime a celebration of effort and blessing rather than just another part of the day. This connection deepens our appreciation for nature's cycles and the importance of sustainable living, aligning closely with principles of stewardship found in scriptures such as Genesis 2:15, where humans are called to "till and keep" the earth.

The Joy Found in Green Spaces

The personal achievement in gardening is palpable. When you watch a plant thrive because of your care, there is a profound sense of accomplishment. This joy is akin to what we feel when we nurture our spiritual lives—growing in faith through acts of love and kindness towards others. Just as we rejoice in spiritual milestones, so too should we celebrate the milestones in our gardens.

This chapter offers practical tips for maximizing these feelings of joy and achievement. Whether it's choosing the right plants for your climate or learning how to optimize your watering schedule, each piece of advice aims to enhance not just the health of your garden but also your enjoyment of it.

Connecting Deeper Through Creation

Gardening invites us to connect deeply with our environment. This connection goes beyond physical interactions with the earth; it encompasses an understanding of how every part of creation is interlinked. When we grow our food, we reduce our carbon footprint, contribute to local wildlife ecosystems, and promote biodiversity—all actions that reflect respect for God's creations.

Moreover, this connection teaches patience and humility—virtues that are essential in both gardening and faith. As Ecclesiastes 3:1 reminds us, "To everything there is a season," emphasizing that timing is crucial, both in planting seeds and in life's many challenges.

In embracing gardening, we accept an invitation to slow down, consider our impact on the world around us, and act thoughtfully. It's not just about cultivating plants but nurturing a spirit attuned to grace and gratitude.

By fostering this deeper connection through hands-on engagement with nature, we are reminded daily of where our food comes from and how much effort it takes to grow—a realization that makes each meal a little more meaningful.

Conclusion

Each chapter in this journey offers not just guidance on gardening techniques but insights into how these actions reflect deeper spiritual truths. By understanding this connection between gardening and spirituality, we learn to appreciate every aspect of creation more fully. This chapter ensures you have all the tools needed to turn your raised-bed into a thriving oasis—a source of food, joy, and spiritual enrichment.

Celebrate the Empowerment from Growing Your Own Produce

Growing your own produce isn't just about filling your plate; it's about taking charge. Imagine the empowerment that comes from planting a tiny seed and nurturing it into a robust vegetable. This process is a vivid testament to the power we hold to shape and influence our immediate environment.

When you grow your own food, you create a direct connection with what you eat. Each leaf, fruit, or root harvested is a product of careful attention and patience, providing not only nourishment but also a sense of achievement. This connection reinforces the value of self-reliance in an increasingly convenience-driven world.

Think of your garden as your personal canvas, where each plant represents a brushstroke contributing to a larger picture of health and sustainability. The act of tending to your garden teaches you about the intricate balance of nature and the importance of nurturing both plants and oneself.

In today's fast-paced society, slowing down to connect with nature through gardening can be profoundly empowering. It's about more than just food; it's about cultivating life skills and resilience. Witnessing the transformation from seeds to sprouts to harvest can boost confidence and foster an appreciation for the cycles of nature.

Celebrating the empowerment from growing your own produce underscores independence and encourages a proactive approach to health and sustainability.

Foster a Deeper Connection to Your Food and the Environment

The bond between gardener and garden goes beyond simple caretaking; it is an intimate relationship with the earth itself. Every time you plunge your hands into soil, you're connecting with the complex ecosystem that makes growth possible.

Did you know that by engaging in gardening, you're participating in an act of environmental stewardship? Gardens are microcosms of biodiversity, supporting everything from microbes in the soil to pollinators like bees and butterflies. This biodiversity is crucial for sustainable ecosystems, which benefit both our gardens and the wider world.

Imagine if every meal were preceded by a moment's reflection on its origins—considering the soil, water, and care that contributed to each bite. Such mindfulness can transform eating from a routine activity into an ecological act that respects our planet's limits.

Gardens teach us about interdependence and ecological responsibility. By choosing native plants or employing organic methods, gardeners can make decisions that support their local environments rather than strain them. These choices ripple outward, influencing broader ecological health.

The rhythms of gardening—planting in spring, harvesting in summer, turning the soil in fall—align us with natural cycles. In doing so, they offer lessons in patience and timing that resonate far beyond the garden's edges.

Could understanding these connections inspire you to see every meal as part of a larger ecological story?

Revel in the Personal Achievement and Joy of Gardening

There's something inherently joyful about watching something grow—a fact known well by anyone who has tended a garden. Each sprout breaking through the soil is a victory against doubt and impatience, embodying potential coming to life right before our eyes.

This joy is not superficial; it's rooted in deep psychological benefits. Gardening has been shown to reduce stress, improve mood, and even enhance cognitive function. It provides not only physical exercise but also mental relaxation—an often-needed respite from our busy lives.

Consider this: each plant nurtured is like crafting a unique piece of art. The gardener is both artist and scientist, blending creativity with knowledge of botany, weather patterns, and wildlife needs. This blend makes gardening an enrichingly holistic activity that engages body, mind, and spirit.

Moreover, sharing the bounty from your garden can multiply this joy. Offering fruits or vegetables to friends or family members allows them to taste not just flavors but also love made tangible through care-grown produce.

By fostering deep connections with our food sources and embracing personal achievements in gardening, we cultivate more than plants—we grow healthier communities rooted in shared joy and environmental care.

Nurturing Your Garden Oasis

Gardening is more than just a hobby; it is a profound journey of empowerment. Through the simple act of planting seeds and nurturing them to growth, you take an active role in managing your own sustenance. This hands-on approach is not just about producing food but fostering a **deep personal connection** to what you eat and to the natural world around you.

The Joy and Achievement of Gardening

Every tomato, pepper, or herb you harvest is a testimony to your dedication and love for gardening. These moments are not merely acts of cultivation but are spiritual milestones that echo the parables of growth and nurturing found in many scriptures. Just as a gardener nurtures a plant, so are we nurtured by our faith and communities, growing together in a symbiotic relationship with nature.

Empowerment through Self-Sufficiency

Growing your own produce does more than just fill your table—it fills your spirit. There's a unique satisfaction in knowing that through your labor, you've contributed directly to your health and well-being. This act of growing food is deeply empowering, providing not only physical nourishment but also spiritual fulfillment. It reminds us of the blessings of abundance and the importance of being stewards of God's creation.

Connect Deeper with Your Environment

Your garden allows you to witness firsthand the impact of your care and effort, linking directly back to the health of our planet. This connection fosters greater environmental stewardship—an essential virtue in today's world. As you tend your garden, consider it an act of worship; caring for the earth as we are called to do.

Revel in Personal Achievements

Celebrate every small victory in your garden—each is a step towards greater self-reliance and resilience. These achievements build not only confidence but also a sense of peace and contentment in being connected to the cycles of life. They teach patience, persistence, and ultimately, the joy of reaping what you sow.

Gardening is an invitation to slow down, breathe deeply, and cultivate more than just plants—it's about cultivating life itself. Whether you are tending a few pots on a balcony or managing a sprawling backyard plot, remember that each seed planted is a step toward a healthier, more sustainable way of living. Let us keep sowing these seeds of hope and harvest joy from our labors.

CHAPTER 7

Earth's Foundation: The Art of Soil Preparation

In the quiet hum of an early afternoon, Thomas knelt beside his raised garden bed, his fingers brushing against the damp soil. His garden was a labor born from necessity, a small patch of green in the heart of the city that reminded him of his father's vast farm back home. As he pressed a seed into the earth, he remembered how his father would speak to the plants, believing they could sense love and care just like any creature.

Thomas had noticed, however, that despite his efforts and whispered encouragements, his plants were not thriving. Their leaves hinted at yellow rather than bursting in lush green hues. He pondered whether it was the soil's pH levels or perhaps its nutrient composition that was lacking. The thought nagged at him as he watched a robin alight briefly on the fence before darting away.

He recalled a conversation with an old friend, a seasoned gardener who once mentioned how crucial good drainage and soil health were for a garden to flourish. Thomas wondered if perhaps his father's effortless success with crops came not just from his tender words but from his unspoken knowledge of these things.

The air carried a chill that hinted at coming rain. Thomas felt it brush against his cheeks as he worked through each section of the garden bed methodically. His hands moved with purpose, yet his mind wandered to last year's harvest — meager and somewhat disheartening. He wanted more than just survival for his plants; he desired for them to thrive, to provide solace and sustenance as they once did under his father's care.

As he adjusted some stones around the edge for better water runoff — an attempt at improving drainage — he paused and looked around at the towering buildings casting long shadows over his little oasis. The juxtaposition of nature against concrete was stark yet beautiful in its own right.

Could understanding deeper elements like compost makeup or precise pH levels be what he needed to bridge the gap between mere survival and vibrant growth? How much of our past knowledge do we carry forward unconsciously, aiding us when we least expect it?

Unlock the Secrets to a Flourishing Garden: Mastering Soil Preparation

Soil is not just dirt; it's the foundation of your garden. In urban environments, where space is limited and soil quality often poor, raised bed gardening becomes a game-changer. However, the success of these gardens heavily depends on one crucial step: **soil preparation**. Proper soil preparation ensures that your plants have the best environment for growth, leading to a more bountiful and vibrant garden.

The essence of effective soil preparation lies in understanding its components and how they interact with plant life. The first key element is **compost**. Compost enriches the soil by adding essential nutrients and improving its structure, which is vital for plant roots to develop and thrive. This natural amendment promotes a balanced ecosystem within your garden soil, encouraging beneficial microbes that aid in plant growth and disease resistance.

Another critical aspect is **checking the soil pH levels**. Different plants require different pH levels to absorb nutrients effectively. Most vegetables thrive in slightly acidic to neutral soil (pH 6.0-7.0). Testing your soil's pH can prevent nutrient lock-up, a condition where plants cannot access nutrients from the soil, regardless of how rich it might be.

Good drainage is equally important for preventing waterlogged roots, which can lead to root rot and other harmful conditions. Raised beds naturally enhance drainage, but it's essential to ensure that the soil structure supports this feature. Incorporating materials like coarse sand or perlite can improve drainage, creating an environment where water and nutrients are adequately available to your plants' roots.

To create a **nutrient-rich environment**, start by layering your raised bed with quality topsoil mixed with ample amounts of compost. Consider adding aged manure, bone meal, or fish emulsion to provide a slow-releasing supply of nutrients throughout the growing season. This approach not only feeds your plants but also keeps them healthy and robust against pests and diseases.

Understanding these principles is not just about technical gardening skills; it's about nurturing life from the ground up. As you prepare your garden's soil, consider it an act of stewardship towards creation, respecting and enhancing the natural resources God has given us. Scripture reminds us in Genesis 2:15 that we are placed in the garden "to work it and take care of it," reflecting our responsibility towards maintaining creation's beauty and bounty.

By following these guidelines, you will create a fertile sanctuary for your plants to flourish. Remember, every successful garden starts with the soil—it's where all life begins in your garden. With patience and care, you'll see that your efforts in preparing this foundation will result in a healthier garden that brings joy and abundance.

Discovering the Significance of Compost and Soil pH

Compost is often referred to as the gardener's gold, and for good reason. It is a rich, decomposed organic matter that provides essential nutrients to the soil, acting as a natural fertilizer. Incorporating compost into the soil of your raised beds not only enriches it with nutrients but also improves soil structure, enhancing water retention and aeration. This is crucial for plant roots to develop properly and access the nutrients they need.

Imagine your garden's soil as a bustling city. Just as a city needs a strong infrastructure to thrive—roads, bridges, and tunnels—your garden soil requires a robust foundation. Compost serves as this foundation, building up the soil's health and creating pathways that allow roots to travel easily and access water and nutrients.

The pH level of the soil, which measures its acidity or alkalinity, significantly influences plant health. Most plants prefer a pH range of 6.0 to 7.0. If the pH is too low or too high, plants may not be able to absorb some nutrients effectively, even if those nutrients are abundant in the soil. Regularly testing the pH level of your soil ensures that it remains within an optimal range for your plants to flourish.

Balancing your soil's pH can sometimes feel like finding the right temperature in an unfamiliar shower—one slight adjustment can make all the difference between too hot and too cold. By adding materials like lime to raise the pH or sulfur to lower it, you can create an environment where your plants will thrive.

In essence, enriching your raised bed with compost and maintaining an appropriate pH balance is fundamental to fostering a healthy garden ecosystem.

Understanding Drainage's Role in Plant Health

Good drainage is vital for preventing water from pooling at the roots of plants which can lead to root rot and other harmful conditions. In raised beds, ensuring proper drainage is easier than in traditional gardens because you can control the soil composition from the start.

Plants' roots are like lungs; they need air to breathe. Overwatered soil drowns these roots, suffocating them and preventing them from absorbing oxygen. This emphasizes why having well-draining soil is crucial—it allows excess water to flow through while retaining enough moisture to keep the roots hydrated.

One way to ensure good drainage is by incorporating coarse materials like sand or gravel into your soil mix. These materials create spaces within the soil structure that facilitate water flow and prevent compaction. Additionally, raising your garden beds above ground level naturally enhances drainage capabilities.

Using rhetorical devices: Isn't it remarkable how adjusting something as simple as the composition of your soil can profoundly impact plant health? It's like tuning an instrument—each adjustment brings you closer to harmony in your garden.

However, while improving drainage might seem straightforward—mix in some sand, raise the bed—it requires careful consideration of what each plant species needs. Some might require more moisture-retentive soils than others, meaning that blanket solutions might not work across different sections of your garden.

Could adjusting our approach to drainage be the key to unlocking healthier gardens?

Creating a Nutrient-Rich Garden Environment
The Soil Success Framework

To create a nutrient-rich environment in raised garden beds begins with understanding and testing your starting point—the existing soil condition. Testing for both pH levels and nutrient content gives you a baseline from which you can make informed decisions on how best to amend your soil.

Step One: Soil Analysis

Start by collecting samples from various points around your raised bed and send these off for analysis or use an at-home testing kit available at most garden centers. Knowing where you stand in terms of pH and nutrient levels will guide you on what amendments are needed.

Step Two: Choosing Amendments

Based on your analysis results, select appropriate types of compost or other organic matter that will benefit your specific type of soil. For instance, if your test shows nitrogen deficiency, adding coffee grounds or manure-based compost might be beneficial.

Step Three: Local Climate Considerations

Your local climate plays an integral role in determining which amendments will perform best. For example, dry climates might require more moisture-retentive organic matter while wetter areas could benefit from faster-draining material additions.

This framework continues through selecting proper planting strategies based on specific needs (e.g., more sand for better drainage) and ends with maintenance practices such as regular additions of organic matter and periodic re-testing of conditions.

Understanding this framework helps visualize how different components interact within your garden's ecosystem: nutrient-rich compost feeds the plants; appropriate amendments adjust pH levels; local climate dictates specific needs; all underpinned by good drainage ensuring healthy root growth.

Through implementing this descriptive framework model in raising bed gardening:

- We establish equilibrium between various elements contributing towards optimal plant growth conditions.
- We see feedback loops where improved soil condition leads to healthier plants which continue enhancing soil quality through natural processes like leaf decay.
- Stability is achieved when plants thrive without constant intervention after initial adjustments have been made correctly.

By focusing on compost enrichment, proper pH balance adjustments, adequate drainage systems, followed by continual

nurturing practices encapsulated within our descriptive framework model 'The Soil Success Framework', we successfully tie together all three key learning objectives into creating sustainable thriving gardens in urban settings

Soil preparation is not just a task; it's the foundation of a thriving garden. **Proper soil management** is crucial for the health and productivity of your plants. By integrating compost, you provide your soil with essential nutrients and improve its structure, which is vital for root development and water retention.

Checking the soil pH isn't merely about numbers; it's about understanding the language of your garden. This knowledge empowers you to adjust the soil conditions optimally for your plants, ensuring they absorb the maximum nutrients. Remember, just as our bodies thrive in the right environment, so do plants in their ideal soil pH.

Good drainage is equally critical. It prevents waterlogging, which can lead to root diseases and suffocation. This aspect of soil preparation might seem technical, but it's about protecting your garden's health, ensuring that every drop of water is a blessing and not a burden.

Creating a nutrient-rich environment is ultimately an act of stewardship. Like the parable of the sower, we want our seeds to fall on fertile ground where they can grow strong and productive (Matthew 13:1-23). Every step you take to enrich and prepare your soil is a step towards a more fruitful harvest.

Let's approach gardening with patience and faith, nurturing our raised beds with diligence and care. The efforts you put into preparing your soil mirror the love and commitment required in nurturing growth—both in gardens and in life.

Embrace these practices with hope and expectation. The work you do today lays down the roots for tomorrow's harvests. Let this be an act of joy, knowing that from small seeds grow magnificent gardens.

CHAPTER 8

A Delicate Balance: The Wisdom of Watering

In the searing heat of a mid-July afternoon, Thomas knelt beside his fledgling vegetable garden, the back of his neck burning despite the shade cast by his wide-brimmed hat. The soil, dry and cracked like an old painting, seemed to cry out for moisture. He pressed his fingers into the earth, feeling its parched texture. This was not just dirt; it was the foundation of life for his plants, a silent partner in his quest for sustenance.

His thoughts wandered to last year's garden—the tomatoes had wilted and the cucumbers had barely fruited. It was a dismal yield that echoed in the empty jars lined up in his pantry. Determination gripped him this season; he would not let history repeat itself. The advice from an old gardening book resonated in his mind, emphasizing deep and consistent watering to reach the roots rather than superficial sprinklings that barely moistened the surface.

As he adjusted the settings on his irrigation system, aiming for a deeper reach into the soil, he recalled a conversation with Mrs. Alder from next door. She had spoken of her roses and how they thrived under her careful attention to their diverse watering needs. Each plant demanded its unique schedule—roses were early morning lovers while her ferns preferred the coolness of dusk.

The sun dipped lower, casting long shadows across Thomas's yard as he worked methodically. He placed mulch around the base of each plant, conserving the moisture they would need to survive this heatwave. His hands moved with a rhythm born of necessity and care, each motion crafting a better environment for growth.

A sudden rustle in the bushes pulled him from his reverie; a small rabbit emerged, its nose twitching as it eyed Thomas warily before bounding off into the neighboring yard. He smiled faintly—nature had its own plans and surprises.

Thomas stood back and wiped sweat from his brow with a dirt-streaked forearm. Would this new approach be enough? Could he adjust in time to see these plants flourish? As he gazed at his garden bathed in golden hour light, hope mingled with doubt—a gardener's constant companions.

What else might Thomas learn from observing nature's cues that could further enhance not just survival but thriving growth in this patch of earth entrusted to him?

Watering: More Than Just a Routine Task

Watering your garden is not just another item on your to-do list; it is a critical practice that can determine the success or failure of your urban gardening efforts. Every drop of water serves as a vital nutrient to your plants, and how you deliver this water can have profound effects on their health and growth. Understanding the intricate balance required in watering techniques can transform a struggling garden into a thriving ecosystem.

The principle of **consistent and deep-rooted watering** practices stands at the core of successful gardening. It's not just about keeping the soil visibly moist but ensuring that water reaches deep into the roots where it is most needed. This method encourages roots to grow deeper, which in turn makes plants more resilient and less susceptible to drought.

To adapt watering schedules effectively, one must first understand the diverse needs of different plants. Each species and even each variety has its unique requirements for water, influenced by factors such as root depth, growth stage, and the local climate. By tailoring your watering approach to these needs, you can ensure that each plant receives the right amount of hydration without wastage.

Achieving optimal plant health through **precise watering techniques** requires both knowledge and patience. It involves observing your plants regularly, understanding their signs of distress, and responding appropriately. Techniques like using a soil moisture meter or simply feeling the soil can provide accurate guidance on when it's time to water again.

For those who follow faith, consider how scripture teaches us about nurturing and stewardship. Just as we are called to care for one another, so too must we care for the Earth and all its inhabitants, including our plants. This stewardship is expressed in our attentiveness to the needs of our garden, mirroring our attentiveness to our relationships and community duties.

Implementing these watering strategies not only promotes plant health but also conserves water—a precious resource. By using methods such as drip irrigation or soaker hoses, you can target water directly to where it's most needed, reducing evaporation and runoff.

By embracing these practices with diligence and faithfulness, you contribute not only to the health of your garden but also to a more sustainable world. Remember, every small step in improving our gardening practices is a step towards better stewardship of the resources entrusted to us. Let us be diligent in this task, showing love not only through our intentions but through our actions as well.

Appreciate the Impact of Consistent and Deep-Rooted Watering Practices

Water is the lifeblood of a garden, much like how our blood nourishes our bodies. For plants to thrive, not just survive, they need more than just a sprinkle on their leaves; they require a consistent supply of water that reaches deep into the earth to soak their roots. This deep-rooted watering encourages the roots to grow downwards, creating a strong foundation that supports the plant above.

Imagine watering a plant is like filling a deep bucket with many holes at different heights. If you only pour a little water, only the topmost holes let water out, affecting only parts of the soil. But if you fill the bucket consistently to the top, water reaches all holes, ensuring every part of the soil is moistened. This analogy helps us understand why thorough soaking benefits plants more than light sprinkling.

Regular and deep watering schedules do more than just quench a plant's thirst; they help in building its resilience against both drought and disease. When roots receive ample water from deep within the soil, they grow stronger and are better equipped to access nutrients that are often deeper underground. This not only enhances their growth but also their ability to withstand when watering isn't possible.

In dry seasons or hot climates, understanding this can be particularly crucial. Without consistent deep watering, plants might develop shallow roots that make them vulnerable to even short periods without water. In contrast, deeply watered plants have a reserve to draw upon, helping them manage stress better during unfavorable conditions.

The key takeaway here is: Deep and consistent watering creates healthier, more resilient plants.

Adapt Watering Schedules to Meet the Diverse Needs of Your Plants

Every plant species has its unique hydration needs which can vary dramatically from one to another. Succulents such as cacti require minimal water compared to thirstier vegetables like cucumbers or tomatoes. Recognizing and adapting your watering schedule to these needs not only conserves water but also promotes optimal health in your garden.

When considering how often and how much to water, think about someone managing a diet plan for different pets; each animal requires specific food amounts and types at various times for best health. Similarly, each type of plant in your garden needs a particular amount of water at certain intervals.

Plants communicate their needs in various ways - wilting may indicate dehydration, while yellowing might suggest over-watering or poor drainage. Being attentive to these signs and adjusting your watering accordingly is crucial. It's not just about giving them what you think they need; it's about responding to what they tell you.

A practical approach could be setting up zones in your garden according to water needs and installing drip irrigation systems with timers adjusted per zone. This targeted watering ensures that each plant gets exactly what it requires without wastage.

Moreover, altering your watering schedule based on seasonal changes and weather conditions is vital for maintaining plant health throughout the year. Increased heat during summer might necessitate more frequent watering sessions compared to cooler months.

Could understanding each plant's unique thirst be **the key** to unlocking lush growth in your garden?

Achieve Optimal Plant Health and Growth with Precise Watering Techniques

Achieving vibrant health and vigorous growth in your garden pivots significantly on mastering precise watering techniques. Each drop of water should be considered precious because when applied correctly, it can lead to flourishing plants that are both beautiful and productive.

Drip irrigation is an example of precision watering at its best. It delivers water directly to where it's most needed - the roots - minimizing evaporation and waste. It's akin to using a syringe instead of a hose; it's precise, controlled, and efficient.

Frequent shallow watering often leads to weak root systems susceptible to various stresses including pest attacks and diseases. In contrast, less frequent but deeper watering encourages robust root growth which naturally enhances overall plant health and productivity due to improved access to nutrients and moisture from deeper layers of soil.

Mulching around your plants can further refine your watering technique by reducing surface evaporation and keeping the soil moist longer. This practice not only saves water but also keeps roots temperate under extreme weather conditions.

By combining these techniques—adjusting schedules per plant needs, using efficient irrigation methods like drip systems or soaker hoses, and supporting with mulches—you can create an optimally hydrated garden environment where every plant achieves its potential.

Integrating consistent deep-rooted practices with adaptive schedules tailored for diverse plant needs using precise techniques forms the cornerstone for vibrant garden health and prolific growth

Understanding the nuances of watering is not just about keeping your garden alive but ensuring it thrives. The techniques we've discussed are foundational, reflecting a profound respect for nature's own systems and a commitment to mirroring this balance in our urban gardens.

Consistent and deep-rooted watering practices are vital. They encourage plants to develop robust root systems that tap into deeper water reserves and nutrients, making them more resilient and healthy. This approach not only conserves water but also builds a garden that sustains itself more naturally over time.

Adapting your watering schedule to the **diverse needs of your plants** is an act of attentiveness and care, akin to understanding the needs of friends or family members. Each plant has unique requirements, and by tuning into these, you cultivate a garden that is both diverse and harmonious.

Lastly, **precise watering techniques** can significantly enhance plant health and growth. By applying the right amount of water at the right time, you avoid the pitfalls of overwatering or underwatering, which can stress plants and lead to poor growth or disease.

Let us remember that gardening is a reflection of our stewardship of the earth. As we tend to our plants with wisdom and care, we are reminded of the delicate balance required to sustain life. In nurturing our gardens, we nurture ourselves, finding joy and peace in the simple act of helping things grow.

May your gardening efforts be blessed with abundance and joy, reflecting the grace and love inherent in nurturing life. Keep these principles in mind as you water your garden, and watch as it flourishes under your careful hand.

CHAPTER 9

Guardians of the Garden: Natural Pest Management

In the quiet hours of an early spring morning, Martha tiptoed through her garden, the dew still fresh on the blades of grass, glistening under the soft glow of dawn. The air was cool and crisp, filled with the earthy scent of wet soil and nascent blooms. Her hands, weathered from years of tending to her plants, moved deftly between the rows of vegetables and flowers she had come to know as well as old friends.

As she inspected a leaf on her prized tomato plant, her brow furrowed. Tiny holes punctuated the green surface—a clear sign of pests. The sight stirred a familiar concern in her chest; not just for the fate of her tomatoes but for all her plants threatened by these tiny invaders.

Martha remembered last year's battle with aphids and how it had nearly cost her the entire crop. Determined not to relive that loss, she pondered over a conversation she had just last week with Ellen from across the fence. Ellen had spoken fervently about companion planting—how marigolds could deter beetles and how basil could keep thrips at bay.

A robin alighted nearby, pulling a worm from the moist earth before flitting away—a reminder of nature's balance. Martha thought about natural predators and how inviting more birds into her garden might serve as a living pest control squad.

She walked back towards her house, past the lavender bushes buzzing softly with early bees. Inside, she pulled out an old notebook crammed with years' worth of gardening notes and sketches. Flipping through it felt like walking down memory lane—every page a testament to trials and triumphs in organic gardening.

Martha sketched a new plan next to notes from last season: marigolds bordering tomatoes, basil alongside bell peppers. She envisioned her garden not just as a source of food but as a vibrant ecosystem where every plant and creature played its role—an interwoven community thriving without harmful chemicals.

As she wrote, Martha's thoughts drifted to her grandchildren playing among rows rich with food and flowers—a future generation learning respect for nature's methods right here in this garden. Would they understand why she chose this path? Would they see how every choice connected like threads in nature's delicate web?

What does it truly mean to nurture life in all its forms within our own backyards?

Unveiling the Secrets to Organic Pest Control

Understanding the challenges of pest management is crucial for every gardener, particularly those committed to maintaining an organic garden in urban environments. As urban gardeners, we are stewards of our little patches of earth, tasked with nurturing them in ways that promote health and sustainability. It's not just about avoiding chemicals—it's about fostering a garden ecosystem that thrives naturally.

The essence of organic gardening lies in the harmony between the plants and their environment. This chapter delves deep into **sustainable pest management strategies** that respect and enhance this balance. By learning to identify common pests and understanding their impact, gardeners can implement effective control measures that do not rely on synthetic chemicals.

Companion planting is one such measure. It's not just about aesthetics; it's a strategic placement of plants which naturally repel pests or attract beneficial insects. For instance, marigolds repel nematodes and other pests while attracting pollinators necessary for garden health. This method not only minimizes pest issues but also enhances the overall productivity of your garden space.

Incorporating *natural predators* is another cornerstone of organic pest control. Introducing or encouraging beneficial insects like ladybugs, which consume aphids, can help manage pest populations effectively. This approach works with nature's own mechanisms to maintain balance and reduces the need for human intervention.

Avoiding harmful chemicals is perhaps the most important lesson in this chapter. The use of synthetic pesticides can lead to a variety of problems, including resistance in pests, harm to non-target species (such as bees), and potential health risks to humans. The philosophy here aligns closely with the principles found in scriptures that emphasize stewardship and care for creation, reminding us that every action we take impacts our environment and our relationship with it.

This chapter aims not only to educate but also to empower you with practical tips on how to implement these strategies in your own urban garden. Whether you're dealing with a small raised bed or a larger community plot, these techniques are scalable and rooted in deep respect for the natural world.

The journey towards mastering organic pest control is both rewarding and challenging. It requires patience, observation, and a willingness to learn continuously about the complexities of nature. But the rewards—safe, healthy produce and a flourishing garden—are immense. Let this knowledge serve as a tool to help you cultivate a space that embodies health, sustainability, and respect for the divine balance of nature.

Sustainable Pest Management in Organic Gardening

Organic gardening thrives on the principle of harmony with nature, emphasizing the importance of maintaining a balance that supports both plant growth and environmental health. A crucial aspect of this balance is pest management, which, if handled sustainably, can enhance both plant health and biodiversity. Sustainable pest management involves methods that are not only effective but also ensure the long-term health of the garden ecosystem.

One foundational method is **crop rotation**. This simple practice involves alternating the types of crops planted in a specific area with each growing season. It disrupts the life cycle of pests that thrive on specific plants, thereby reducing their chance to multiply. By rotating crops, gardeners can naturally reduce pest populations without the need for chemical interventions.

Imagine your garden as a dance floor, where each type of plant has its own rhythm and pests are uninvited guests who favor certain tunes. Crop rotation changes the music periodically, discouraging these pests from settling in. It's a natural strategy to keep pests guessing and away from your plants.

Another sustainable practice is the use of **physical barriers**, such as nets or floating row covers, which can protect plants from insects and birds without harming them or the environment. These barriers act like security guards, shielding your plants from invaders looking to feast on them. They are especially useful for vulnerable young seedlings or soft fruits that attract birds.

Biological control is yet another layer in sustainable pest management. Introducing or encouraging beneficial insects that prey on harmful pests can maintain ecological balance and reduce reliance on chemical pesticides. For instance, ladybugs devour aphids, and nematodes can tackle soil-borne larvae.

Boldly integrating these sustainable methods into your gardening practices ensures a thriving garden ecosystem where every plant and creature plays its part in maintaining balance.

Companion Planting and Natural Predators: A Symphony of Interactions

Companion planting is not merely placing different plants next to each other; it's about creating a community where each member supports and enhances the growth of others. Marigolds, for example, emit a natural substance that deters many garden pests and even some harmful nematodes in the soil. This makes them excellent companions for vegetables like tomatoes or peppers, which might otherwise fall victim to these underground attacks.

The concept extends beyond mere repulsion of pests. Some plants, such as basil, can improve the flavor and growth rate of their garden neighbors like tomatoes. This synergistic effect not only maximizes your garden space but also enhances overall plant health, reducing susceptibility to pests.

But how does one harness this natural potential? Start by understanding which plants complement each other and plan your garden layout accordingly. Planting garlic near roses can help deter aphids; similarly, tall sunflowers can provide shade needed by lower-growing lettuce in hot climates.

Incorporating **natural predators** forms another pillar in this organic strategy against pests. Introducing or attracting predators such as birds, frogs, or beneficial insects like bees into your garden creates a natural defense system. These predators provide critical control over pest populations naturally.

Consider this: if your garden were an orchestra, then every plant and predator plays an instrument, contributing to the harmony of your garden's symphony against pests.

Could understanding these interactions be the key to unlocking healthier gardens without chemicals?

The Dangers of Chemical Pesticides: Choosing Health Over Harm

Chemical pesticides offer quick fixes to pest problems but at a considerable cost to both environmental health and human safety. These chemicals often don't discriminate between harmful pests and beneficial organisms; they kill indiscriminately, disrupting ecological balances and leading to cycles of dependency on chemical solutions.

Moreover, many pesticides have been linked to serious health issues in humans—from headaches and skin irritations to more severe implications like respiratory problems or even cancer with long-term exposure. Their residue can linger in soil and waterways long after they have been applied, posing ongoing risks.

Consider instead a gardener who chooses natural methods over synthetic chemicals—they are like someone opting for nutritious food over fast food; it requires more knowledge and patience but results in better health and well-being.

Educating oneself about **the dangers associated with chemical pesticides** is essential for making informed decisions about their use in gardens. Opting for safer alternatives not only protects your health but also contributes to a healthier planet.

By integrating sustainable pest management practices, utilizing companion planting strategies coupled with natural predators, gardeners create robust systems that maintain high yields without chemical reliance.

Understanding these environmentally friendly approaches empowers you to create lush gardens that flourish naturally

while safeguarding both plant life and human health—achieving true guardianship over your green spaces.

Mastering the art of natural pest management is not just a gardening skill—it's a commitment to stewardship of the earth God entrusted to us. By embracing sustainable methods, we align our actions with the wisdom found in Scripture, where we are called to care for creation responsibly and lovingly. This approach not only preserves the health of our gardens but also protects the broader environment and its creatures.

Sustainable pest management is essential for maintaining an organic garden that thrives without the intervention of harsh chemicals. These substances can harm more than just pests; they affect our soil, water, and even our own health. As gardeners, it's imperative to seek alternatives that coincide with the natural world's rhythm, mirroring the balance and harmony designed by our Creator.

Utilizing techniques like **companion planting and encouraging natural predators** proves not only effective but also enriches our understanding of ecological relationships. These methods illustrate beautifully how interconnected God's creation is, with each plant and animal playing a role in a larger system. By enhancing biodiversity in our gardens, we encourage a healthy ecosystem that naturally keeps pest populations in check.

Moreover, avoiding harmful chemicals is not merely a practical gardening choice but a moral one as well. It reflects our respect for life and our desire to pass on a healthier, more vibrant world to future generations. In Galatians 6:7, we are reminded, "Do not be deceived: God cannot be mocked. A man reaps what he sows."

When we sow care and diligence in our gardens, avoiding substances that bring harm, we reap a harvest that is both plentiful and wholesome.

In your gardening journey, remember that each small step towards sustainable pest management is a leap towards fulfilling a greater responsibility. It's an act of faith in practice—a faith that believes in nurturing and sustaining rather than dominating and depleting.

May your garden be a testament to your care for God's creation, reflecting the beauty and abundance He intended. Let it be a space where nature's resilience and diversity are celebrated, where each plant thrives because of your commitment to gentle stewardship. Continue learning, experimenting with natural solutions, and sharing your experiences with others. Together, as a community of gardeners rooted in faith and love for the earth, we can make a significant impact.

Your efforts are not only growing plants but also growing a legacy of environmental empathy and action—an invaluable contribution to our world.

CHAPTER 10

The Circle of Life: Sustainable & Healthy Gardening

In the soft glow of the late afternoon, Thomas walked through his modest garden, the soil cool and yielding beneath his boots. His hands, worn from years of tilling and planting, hovered over the young tomato plants that were just beginning to show their promising fruits. The air was filled with the scent of basil and earth—a simple pleasure that he cherished deeply.

As he knelt beside a row of carrots, his mind wandered to the council meeting scheduled for next week. The town was considering a new proposal to expand industrial farming on the outskirts, a plan that weighed heavily on him. He feared for the local ecosystems, for the bees that buzzed tirelessly around his lavender bushes, and for his own small patch of greenery that might soon be overshadowed by vast fields treated with chemicals.

Thomas remembered his grandfather's stories of old farming practices, how they used to rotate crops and let fields rest. It seemed such knowledge was slipping away in favor of quicker yields. He pulled a weed from around a pepper plant gently; it came loose with little resistance.

A neighbor's laughter floated over the fence, snapping Thomas back from his reverie. Emily stood there, her young son tugging at her skirt as they peered curiously into Thomas's garden. "Your veggies always look so vibrant, Thomas," Emily called out. "What's your secret?"

"It's no secret really," Thomas replied with a smile as he stood up, brushing dirt from his knees. "Just good old-fashioned organic methods; taking care of the earth so it takes care of us."

Emily nodded thoughtfully as her son watched a butterfly flit by. She seemed concerned yet hopeful—a reflection of Thomas's own feelings about organic gardening in these changing times.

As he waved goodbye and turned back to his garden, Thomas thought about how every choice he made here connected to something much larger than himself. Could simple acts like choosing sustainable gardening practices ripple outwards far enough to make real change?

Was it possible for one small garden to help turn the tide against larger environmental challenges?

Nurturing Growth: How Your Garden Can Change the World

Gardening is more than just a hobby; it's a powerful act of stewardship. As urban gardeners, we have a unique opportunity to influence both our immediate environment and the broader ecological landscape. This chapter delves into how adopting sustainable and healthy gardening practices can create a ripple effect, enhancing personal well-being and contributing to global health.

Embracing organic gardening is not merely about avoiding chemicals, but nurturing the soil and everything that springs from it. Organic practices ensure that the food we grow is not only safe for consumption but also richer in nutrients. This approach echoes the biblical principle of keeping the Earth fruitful and multiplying its yield without harm (Genesis 1:22). By choosing organic methods, gardeners make a conscious decision to protect their health and that of their communities.

Understanding the **role of sustainable methods** in gardening goes beyond knowing what to plant and when. It involves recognizing the interconnectedness of all living things and how our choices impact this delicate balance. Techniques such as crop rotation, composting, and the use of natural pest deterrents contribute not only to a thriving garden but also support surrounding wildlife and reduce carbon footprints. These actions reflect the wisdom of living harmoniously with nature, as advised in Leviticus 25:4, which speaks to letting the land rest.

Reflecting on our gardening choices allows us to see **the broader impact on global ecology**. Each decision, from conserving water with drought-resistant plants to creating habitats for pollinators, plays a part in a much larger narrative. As stewards of Earth, it's crucial we model responsibility for future generations, embodying stewardship as seen in Matthew 25:21, where faithful management is rewarded.

To truly integrate these practices into your gardening routine, start small. Choose one new organic technique each season or research native plants that naturally enhance your garden's sustainability. Share your journey with neighbors or online communities to inspire others and build a network of eco-conscious gardeners.

The benefits of sustainable gardening extend far beyond the confines of your own backyard. They reach into the very fabric of community health and global environmental management. By choosing this path, you actively participate in a movement that values life in all its forms.

Remember, every plant in your garden can be a testament to care and consideration for creation. Let your garden be a place where life thrives in its full glory—healthy, sustainable, and abundant.

Embracing Organic Gardening Practices

Organic gardening is not just a method; it's a commitment to fostering the health of both our environment and ourselves. By avoiding synthetic chemicals and pesticides, organic gardeners ensure that the food they grow is not only nutritious but also free of harmful substances. This practice supports biodiversity by maintaining healthy soils and ecosystems, which in turn supports more vibrant plant life.

Imagine your garden as a teacup. Just as you would not pour harmful chemicals into your tea, treating your garden similarly ensures that everything growing within it is nurtured by nature's own mechanisms. Composting, companion planting, and natural pest control are pillars of organic gardening that help maintain this balance.

The benefits of organic gardening extend beyond the borders of your own yard. Reducing chemical runoff protects local waterways and wildlife, which are often the first victims of environmental mismanagement. Furthermore, organic gardens can act as mini carbon sinks, absorbing carbon dioxide from the atmosphere and helping to mitigate climate change.

Organic methods also encourage gardeners to think creatively about resources. Rainwater harvesting and solar-powered garden lights are examples of how one can maintain a garden sustainably. These practices not only reduce reliance on non-renewable resources but also foster a deeper connection between the gardener and their environment.

Embracing organic gardening helps preserve our planet while providing us with pure, healthy produce.

Understanding Sustainable Methods in Gardening

Sustainable gardening is integral to long-term ecological balance and resource efficiency. It involves techniques that do not deplete resources or cause long-term damage to the environment. By implementing sustainable practices, gardeners can create systems that self-maintain, requiring less input of water, fertilizers, and other resources over time.

Water conservation strategies like drip irrigation or xeriscaping help reduce the wasteful use of water and ensure that plants receive just what they need. Such methods not only save water but also teach us about efficiency and moderation in resource use.

Sustainability in gardening also involves choosing native plants which are adapted to local conditions. These plants require less water, fewer fertilizers, and minimal care compared to non-native species that might struggle or become invasive. This selection process mirrors how nature itself evolves—favoring species that thrive in their inherent conditions without external aid.

Using renewable energy sources in the garden—such as solar panels to power greenhouse fans or water features—aligns with sustainable living principles by reducing fossil fuel dependence. This approach exemplifies how every small choice in favor of sustainability contributes to global environmental health.

The introduction of organic waste recycling through composting transforms waste into wealth by turning kitchen scraps and yard debris into nutrient-rich compost. This cycle mimics nature's own recycling system where nothing goes to waste.

Could recognizing the impact of each sustainable choice in your garden inspire you to rethink other areas of resource use in your life?

Reflecting on Broader Global Impacts

Every decision made in our gardens ripples across the globe in ways we might not immediately realize. Choosing to buy peat-free soil or opting for seeds from heirloom varieties are decisions that support global biodiversity and prevent habitat destruction.

The analogy here could be as simple as viewing each garden as a single note in a grand symphony—the collective effect of these individual choices creates a harmony that either supports or degrades our global ecosystem. When we choose sustainable methods, we're composing a melody that respects and enriches all life on Earth.

By fostering biodiversity through planting diverse species and creating habitats for wildlife, gardens become microcosms of larger ecological networks. This biodiversity is crucial for resilience against pests, diseases, and climatic changes—challenges that are increasingly impacting us on a global scale.

Moreover, educating ourselves about the origins of our gardening supplies—such as where our seeds come from or how our garden tools are made—can make us more conscious consumers who support ethical businesses and fair labor practices worldwide.

Reflecting on our gardening choices allows us to see ourselves as partakers and protectors of Earth's vast ecosystem; it encourages us to foster sustainability not just in our backyards but across all areas of our lives.

Embracing Organic Practices

Organic gardening is not just a method; it's a commitment to fostering health and well-being both in our gardens and within ourselves. By choosing to avoid synthetic chemicals, we protect the integrity of our food and the safety of our families. This choice reverberates beyond our personal spaces, contributing to broader environmental health. Scripture teaches us about stewardship, about caring for the Earth that has been entrusted to us (Genesis 2:15). In practicing organic gardening, we honor this divine responsibility, ensuring that our land remains fruitful and unspoiled for future generations.

Sustainable Methods: A Cornerstone for Success

Sustainability in gardening goes beyond mere plant survival. It encompasses water conservation, soil enrichment, and the wise use of resources. These practices are not only practical but are reflections of the principles found in Matthew 25:21, where faithful stewardship is rewarded. By implementing sustainable techniques, gardeners create systems that are not just self-sufficient but resilient and productive over time. The success of these methods lies in their ability to support continuous growth, mirroring the perpetual nourishment provided by God's creation.

Reflecting on Our Global Impact

Every choice we make in our garden—be it the type of seeds we sow or the methods we employ—casts ripples across the global ecological pond. When we choose paths that harm less and replenish more, we align ourselves with God's call to love and care for all creation (Psalm 24:1). This reflection is crucial as it guides us to make decisions that uphold the sanctity of life and promote a balance that sustains all living beings.

In embracing these practices, we do more than grow food—we cultivate hope, health, and harmony within our environment. This harmonious approach is essential not only for our immediate surroundings but for the global community as well. As stewards of Earth, it is both a privilege and responsibility to ensure that our actions contribute positively to the world around us.

Let us move forward with faith and diligence, nurturing our gardens as we nurture our souls—mindful always of the profound impact our choices have on this planet God has so lovingly created. Through mindful gardening, we live out the love shown through Christ—caring profoundly for every part of creation entrusted to us.

CHAPTER 11

Grow Where You're Planted: Selecting the Right Plants

In the heart of a small Midwestern town, where the soil was as rich in stories as it was in nutrients, Sarah stood in her modest garden. The sun dipped low, casting long shadows that danced between the rows of budding plants and vegetables. She wiped a bead of sweat from her brow, her hands stained with earth, evidence of a day spent tending to the life burgeoning under her care.

Her mind wandered to the advice she had received from an old gardener at the market last week. "Understand your zone," he had said with a knowing smile, his voice rough like gravel yet comforting. It wasn't just about planting seeds; it was about knowing where they could thrive. Sarah pondered over this as she looked at her garden, considering how each plant's success was tied to the invisible lines of a grower's map.

A gentle breeze stirred, bringing with it the scent of jasmine from Mrs. Henderson's yard next door. It reminded Sarah of springtimes past when her own attempts at gardening were less informed — hopeful endeavors met with the harsh reality of frost and scorching sun alike.

She recalled her failures vividly: tomatoes that wouldn't set fruit because the nights were too cool, lavender that wilted under the oppressive heat of midsummer. Each loss was a quiet echo in her heart, a reminder of what could have been if only she had known more about the delicate balance needed for each species.

Now armed with knowledge and a newfound respect for growing zones and environmental conditions specific to her region, she felt more confident. The local extension office had provided maps and guides that detailed which plants would flourish and which would falter in zone 5b.

As she adjusted a young maple sapling that seemed to lean too far towards the setting sun, Sarah thought about how choosing suitable plants for one's garden was much like selecting people in one's life — not everyone would thrive in just any condition; they needed an environment suited to their growth.

A robin landed near by, tilting its head curiously as if inspecting Sarah's work or perhaps reflecting on its own instinctual knowledge of survival and adaptation. She smiled at it momentarily before returning to her thoughts.

Would understanding these invisible boundaries help not just her garden but also herself grow?

Thrive in Your Zone: The Key to Garden Success

Navigating the significance of growing zones in plant selection is more than just a gardening tip; it's a fundamental principle that can make or break your gardening efforts. Understanding these zones helps gardeners recognize the unique climate conditions of their area, which directly influences which plants will flourish. Every plant has specific needs, and when those needs align with the environmental conditions you can provide, you set the stage for a thriving garden.

Growing zones, defined by variables like temperature, humidity, and sunlight hours, serve as a guide to help you choose plants suited to your specific environmental conditions. This isn't just about ensuring your plants survive; it's about helping them thrive. For instance, planting a species that prefers arid climates in a region that experiences heavy rainfall might not yield successful results. Conversely, selecting plants that are adapted to your local conditions increases your garden's success rate by reducing stress on the plants and minimizing the need for extra resources like water and fertilizers.

This approach not only makes your gardening more sustainable but also more rewarding. By selecting appropriate plant varieties, you're not just growing plants; you're nurturing a relationship with your environment. This harmonious interaction between plant and place does more than produce a bountiful harvest; it creates a sense of connection and stewardship towards nature.

For urban gardeners working with limited space like raised beds, this knowledge becomes even more crucial. Raised bed gardening offers excellent soil control, which can be an advantage if you know how to leverage it with the right plant choices based on your growing zone. Such strategic planting ensures that every square inch of your garden is optimized for production and beauty.

Harness Your Environment: A Guide to Informed Plant Selection

The process begins by identifying your growing zone using available resources such as USDA zone maps or local gardening clubs. Once you know your zone, choosing plants becomes an informed decision rather than a gamble. This is where faith in the natural order—understanding that every plant was designed with a purpose and ideal setting—can guide us to make choices that respect and enhance the natural ecosystem.

Moreover, embracing the constraints of your local environment doesn't limit you; it frees you to explore a diverse array of plants ideally suited for your conditions. This tailored approach reduces common gardening frustrations such as pest infestations, diseases, and poor yields, which often result from environmental mismatches.

In practical terms, when selecting plants for your garden, consider not only the aesthetics but also their compatibility with your climate. Research plants that share similar needs in terms of light exposure, water requirements, and soil type as those native to or thriving in your zone. For example, Mediterranean herbs like rosemary and thyme perform well in hot, dry zones while ferns and hostas may be better choices for shadier, moister areas.

Cultivate Success Through Strategic Planting

Ultimately, understanding and utilizing growing zones is about cultivating a garden that is resilient and abundant. It's about making wise choices today that will lead to less heartache tomorrow. As stewards of our little patches of earth—no matter how urbanized—we have the opportunity to create pockets of biodiversity that reflect both our personal tastes and our ecological responsibilities.

Embrace this journey with patience and faith, knowing that each small decision aligns with greater natural rhythms. Your efforts are not only creating beauty but also contributing to the well-being of our planet by fostering environments where life can flourish naturally. This approach makes gardening an act of love—a way to give back to the land that sustains us—and reminds us of our place within the larger tapestry of life.

Navigating the Significance of Growing Zones

Understanding growing zones is akin to knowing the dress code for an event; it ensures you're appropriately prepared and can thrive in the given environment. Growing zones, officially known as Plant Hardiness Zones, are geographically defined areas depicted on a map, designed by the USDA, that signify where various perennial plants are most likely to thrive based on climatic conditions. These zones are primarily determined by the average annual minimum winter temperature and provide a useful guide for landscapers and gardeners.

Each zone differs slightly in terms of temperature ranges, which can greatly influence the survival and growth rate of plants. For instance, a plant that prospers in the warm climes of zone 9 might find it challenging to survive the harsh winters of zone 4. This fundamental understanding helps prevent mismatches between plants and environments, reducing plant mortality rates and enhancing garden success.

Imagine planting tropical hibiscus in a snowy city; they would shiver and struggle like a beachgoer in Antarctica! This metaphor underscores the importance of aligning your plant choices with the climate specifics of your growing zone. By doing so, you not only safeguard your plant's health but also optimize your garden's overall productivity.

To further refine your gardening strategy, consider microclimates within your local area that might affect plant growth. Microclimates are small areas where the climate differs from the surrounding area – perhaps due to elevation, proximity to water bodies, or urban structures – and can influence what plants will best suit a specific part of your garden.

Understanding and respecting growing zones is essential for selecting plants that will flourish in your garden.

Choosing Plants Suited to Your Specific Environmental Conditions

Selecting plants that are suited to your local environmental conditions goes beyond just knowing the hardiness zone; it involves a comprehensive assessment of soil type, sunlight exposure, humidity levels, and more. Each of these factors plays a crucial role in ensuring that your chosen plants not only survive but thrive.

Soil type, for instance, can vary dramatically within the same geographical area and significantly affects water drainage and nutrient availability. Some plants may prefer well-drained sandy soil while others thrive in moisture-retentive clay. Similarly, understanding the sunlight patterns throughout your garden is crucial. While some plants require full sun to bloom abundantly, others might need partial shade to protect them from burning.

Incorporating knowledge about local pests and diseases into your plant selection process can save you a lot of trouble down the line. Plants native to your region are generally more resistant to local pests and diseases and will require less maintenance.

Consider this analogy: just as you would choose footwear based on both fit and function – sneakers for running, boots for hiking – so too should you select plants based on both their aesthetic appeal and their suitability for your garden's specific conditions.

By thoughtfully selecting plants adapted to their environment, gardeners can create a more sustainable and low-maintenance garden. This proactive approach minimizes resource wastage – such as water, fertilizers, and time – ultimately making gardening more enjoyable and rewarding.

Could understanding every layer of your garden's environment lead you to create a perfect green oasis?

Increasing Your Garden's Success Rate with Appropriate Plant Varieties

Choosing the right varieties isn't just about aesthetics; it's about survival and sustainability. When selecting plants for your garden, consider their compatibility not only with the climate but also with each other. Some plants can enhance each other's growth when planted together — a concept known as companion planting — while others might inhibit each other due to competition for resources.

For example, tomatoes thrive when planted alongside basil because basil helps repel harmful insects naturally. This strategic pairing increases yield without resorting to chemical pesticides. Such informed decisions elevate a simple garden into an ecosystem where plant relationships contribute to mutual success.

Further increasing your garden's success rate involves regular monitoring and adaptation. Gardens are dynamic environments; what works one year might not work another due to changes in weather patterns or soil conditions. Being observant and responsive to these changes makes gardening an ongoing learning process.

A good gardener understands that like tuning an instrument, cultivating a garden requires continuous adjustment and fine-tuning based on feedback from the plants themselves.

By carefully selecting appropriate plant varieties tailored to specific growing conditions and interactions within the garden ecosystem, you maximize both health and yield in your gardening endeavors.

Gardening successfully requires recognizing growing zones' significance, choosing environmentally suited plants, and selecting harmonious varieties — all integral steps toward cultivating a thriving garden space

Understanding how to select the right plants for your garden begins with a deep appreciation of **growing zones**. These zones guide us not merely in gardening but in stewardship of our environment, reflecting a principle we find in scripture: "The earth is the Lord's, and everything in it" (Psalm 24:1). Recognizing this, we see that each plant placed in its right zone is an act of aligning with God's natural design.

Choosing plants that match your specific environmental conditions goes beyond mere aesthetics or yield. It's about creating a harmonious space where every living thing can thrive — a concept that resonates deeply with the biblical call to cultivate and keep the earth (Genesis 2:15). This stewardship is not only a duty but a privilege, allowing us to witness the interconnectedness of God's creation directly in our backyards.

Moreover, selecting appropriate plant varieties enhances the likelihood of a thriving garden. This mirrors the wisdom found in Ecclesiastes 3:1, "There is a time for everything, and a season for every activity under the heavens." Just as there is a season for every purpose, there is a variety suited for every garden. By choosing wisely, we increase our garden's success rate and fulfill our role as caretakers of our given environment.

Let us approach gardening with joy and reverence. As you plant each seed or sapling, remember that you are participating in an ancient cycle of life, one that offers not only sustenance but also spiritual reflection and growth. Embrace this task with faith and patience, knowing that your labor is both an act of love and a reflection of divine grace.

In nurturing our gardens, we nurture ourselves and deepen our connection with creation. Every choice made at the gardening store or nursery can be seen as an echo of divine wisdom. Let these choices be made thoughtfully and prayerfully, ensuring that each plant not only grows but flourishes where it has been planted.

By adhering to these principles, you set the stage for a fruitful harvest and a deeper appreciation of the natural world. May your garden be a source of beauty and inspiration, reminding you always of the boundless creativity of our Creator.

CHAPTER 12

A Crop in Time: The Genius of Rotation

Amid the gentle undulations of a vast farmland in Iowa, under the vast dome of a mid-afternoon sky, Thomas maneuvered his old tractor across the field, its wheels churning up the rich earth. The scent of freshly turned soil mingled with a faint whiff of diesel, a perfume that spoke of years tending to these lands. His hands, weathered like the leather seat he sat on, gripped the steering wheel with an ease born of routine.

In his mind, Thomas replayed conversations about crop rotation—a practice his father had sworn by and which he, in turn, had adopted. He remembered his father's voice, rough as bark, explaining how rotating corn with soybeans helped keep the soil healthy. "It's all about giving back what we take," he'd said. Now, as Thomas watched crows scatter from his path, he wondered if there was more he could do. The challenge was no longer just about yield—it was about sustainability.

As he paused at the end of a row to wipe sweat from his brow with a faded blue handkerchief, Thomas looked over to the neighboring field where Janet worked. Her farm had started experimenting with more diverse rotations—adding oats and alfalfa into the mix—and her enthusiasm was both infectious and intimidating. She claimed it cut down on pests naturally and even improved her yields.

The idea nagged at him as he restarted the engine. Could introducing more variety be the key? He thought about his own fields—monotonous stretches of just two crops—and felt a twinge of doubt. Was he being stubborn by clinging to old methods when new techniques could potentially benefit his soil and crop health?

His tractor rumbled on and so did his thoughts; they were interrupted only by an unexpected visit from Mark from two farms over. As Mark leaned against the tractor wheel with a friendly grin discussing rainfall patterns this season, Thomas listened but partway through found himself sharing his concerns about pest management and soil depletion.

Mark nodded thoughtfully as Thomas spoke of Janet's success. "Makes sense," Mark mused aloud. "Different plants pull different nutrients; rotate 'em right and you replenish your soil instead of stripping it bare."

As they parted ways with promises to catch up again soon at Helen's diner in town—a place peppered with tales like theirs—Thomas felt a flicker of resolve stir within him.

Driving back home under an amber wash of sunset that streaked across vast fields turning them gold for moments too brief, Thomas mulled over everything once again: His father's lessons rooted in tradition versus Janet's spirited innovations fueled by hope for tomorrow.

Could embracing these new strategies be what it takes to sustain not only his farm but also a legacy? How does one balance respect for past wisdoms while courting uncharted successes promised by change?

Unlocking the Secrets of Soil: The Power of Rotation

The ancient practice of crop rotation is not just a relic of agricultural history; it is a vital strategy for today's urban gardeners who rely on raised bed gardens. This technique, which involves alternating the types of crops grown in the same space with each passing season, plays a crucial role in maintaining soil health, managing nutrients, and combating pest and disease outbreaks.

Understanding the benefits of crop rotation for soil health and nutrient management is fundamental. In simple terms, different plants require different nutrients and in varying amounts. When the same crop is planted repeatedly in the same soil, specific nutrients can be depleted, leading to poor plant health and reduced yields. By rotating crops, you ensure that the soil's nutrient balance is maintained, as different crops absorb and replenish varied nutrients. For example, legumes can fix nitrogen from the air, enriching the soil for the next crop that might be more nitrogen-demanding.

Moreover, **reducing pest and disease risks through strategic plant rotation** cannot be overstated. Many pests and diseases tend to target specific plant families. If crops are not rotated, these pests and pathogens can build up in the soil, ready to attack the next season's plants. Rotating away from a specific plant family can disrupt the life cycles of these pests and diseases, significantly reducing their impact.

Enhancing soil fertility and overall plant health over successive seasons is another key lesson to embrace. Crop rotation fosters a diverse microbial environment which is essential for a healthy soil ecosystem. These microbes play a role in breaking down organic matter into nutrients that plants can easily absorb, improving soil structure and fertility over time.

For urban gardeners working with limited space in raised beds, this practice is especially important. It allows you to maximize your use of space while ensuring that your small-scale garden remains productive and vibrant year after year. It teaches patience and planning but rewards you with a more resilient garden.

Adopting crop rotation isn't just about following a set pattern; it's about understanding your plants' needs and how they interact with their environment. This knowledge empowers you to make informed decisions that boost your garden's productivity and sustainability.

Implementing these strategies may require some initial adjustment and learning. However, remember that every seasoned gardener was once a beginner too. Start simple—with rotating just a few crops—and gradually expand as you gain confidence and observe positive changes in your garden's health.

By integrating these practices into your gardening routine, you're not only enhancing your own garden but also contributing positively to your local ecosystem. Embrace the challenge of learning about different plants' needs and enjoy the process of nurturing your garden through thoughtful rotation practices. Your dedication will certainly yield flourishing results season after season.

Understanding the Benefits of Crop Rotation for Soil Health and Nutrient Management

Crop rotation, the practice of growing different types of crops in the same area across a sequence of seasons, is not just a farming technique but a cornerstone of sustainable gardening. This method helps in maintaining the health and fertility of the soil, which is crucial in a raised-bed garden where soil depth and volume are limited. By alternating crops that have varying nutrient needs and contributions, gardeners can naturally manage soil fertility without relying heavily on chemical fertilizers.

Imagine your garden as a bustling café, where each type of plant orders a different meal—nutrients from the soil. Just like how repeatedly serving the same meal can deplete the kitchen's supplies, growing the same crop repeatedly can deplete specific nutrients from the soil. Crop rotation is akin to changing the menu regularly to ensure that one type of meal doesn't exhaust all ingredients.

Different plants contribute differently to soil health. Legumes, for example, fix nitrogen from the atmosphere and add it to the soil, benefiting subsequent crops that might be more demanding in nitrogen, such as leafy greens or corn. On the other hand, deep-rooted vegetables like carrots can help bring up nutrients from deeper layers of the soil, making them available to plants with shallower roots.

The benefits extend beyond nutrient management. Rotating crops also helps in breaking cycles of pests and diseases associated with specific plants. If a pest thrives on a particular plant, not planting that crop again immediately in the same spot can disrupt the life cycle of the pest, reducing its chances of survival.

By practicing crop rotation, gardeners optimize nutrient use within the soil, ensuring each plant gets what it needs without depleting the earth's resources.

Reducing Pest and Disease Risks Through Strategic Plant Rotation

Crop rotation plays a pivotal role in managing pests and diseases in your garden. Many pests and diseases are host-specific, meaning they prefer certain types of plants over others. By rotating your crops—changing what you plant where—you effectively disrupt these pests' habitat and life cycles.

Consider this analogy: If pests are like unwelcome house guests who only eat certain types of food, rotating crops is like removing their favorite dishes from the menu each season. This strategy forces them to move on or starve, thus protecting your new crops from becoming their next meal.

This practice not only deals with underground pests but also with airborne spores from plant diseases such as blight or mildew which can linger after harvest. When you plant a different species in place of one that was previously infected, it reduces the chances that disease will establish itself again since most pathogens target specific plant families.

Additionally, strategic rotation can improve soil structure and reduce weed pressure. Certain crops can suppress weeds through natural chemicals they release or by providing dense foliage coverage which limits weed growth.

Incorporating flowering plants or herbs in rotations can attract beneficial insects that prey on harmful pests. This ecological approach not only reduces your reliance on chemical pesticides but also enhances biodiversity in your garden.

Could changing up your garden's "menu" be the secret sauce to robust plant health?

Enhancing Soil Fertility and Plant Health Over Successive Seasons

The Crop Rotation Analytical Framework (CRAF)

To effectively implement crop rotation in raised-bed gardens while maximizing benefits such as enhanced soil fertility and healthier plants over time, we introduce an analytical framework called CRAF—Crop Rotation Analytical Framework.

Nutrient Demand Categorization

Firstly, CRAF categorizes crops based on their nutrient demands: heavy feeders (e.g., tomatoes), light feeders (e.g., onions), and nitrogen fixers (e.g., beans). Understanding which category each crop belongs to helps plan a rotation schedule that balances nutrient depletion with replenishment.

Pest Resistance Analysis

Secondly, it involves analyzing crops for their resistance to common pests and diseases. This analysis helps select crop sequences that naturally reduce pest populations or disease spread without chemical interventions.

Rotational Timing Sequence

The timing and sequence component is crucial—it suggests optimal periods for rotating specific types of crops based on their growing seasons and life cycles. This ensures that each crop has adequate conditions for growth while contributing positively to overall garden health.

Practical Implications

Implementing CRAF means observing how different plants interact with each other and affect various aspects of soil health over time—such as pH levels or organic matter content—and adjusting plans accordingly. For small spaces like raised beds or urban gardens, CRAF recommends vertical cropping or companion planting alongside regular rotations to make efficient use of space while enhancing crop rotation benefits.

This framework enables gardeners to create robust planting schedules that ensure long-term productivity and sustainability in their gardens by balancing immediate crop needs with future soil health considerations.

By understanding these key concepts—nutrient management through diverse planting; disrupting pest habitats through strategic rotations; enhancing long-term soil fertility; and implementing frameworks like CRAF—gardeners can achieve more productive gardens that are also environmentally friendly.

Embracing Rotation for a Thriving Garden

Understanding the powerful role of crop rotation in maintaining **soil health** and enhancing **plant vitality** is essential for every gardener. By wisely rotating crops, you are not just planting seeds; you are sowing the future health of your garden. This practice mirrors the wisdom found in Ecclesiastes 3:1, "To everything there is a season, and a time to every purpose under the heaven." Just as seasons change, so should the crops in your raised beds to ensure each plant contributes to and benefits from the soil's nutrients.

Strategic rotation reduces the risk of pests and diseases that can otherwise thrive if the same family of plants is cultivated repeatedly in the same soil. This proactive approach aligns with the biblical principle of stewardship—taking care of our resources responsibly to reap a bountiful harvest. By rotating crops, you actively participate in preventing potential problems rather than reacting to them after they arise.

Moreover, enhancing soil fertility through rotation ensures that your garden remains productive season after season. It's an investment in the land that pays dividends in the form of robust plants and abundant yields. This method reflects a deep respect for nature's processes and an understanding of how to work harmoniously with them.

Cultivating Patience and Persistence

Implementing crop rotation might require patience and persistence, but remember, "Let us not become weary in doing good, for at the proper time we will reap a harvest if we do not give up" (Galatians 6:9). Each season brings its challenges, yet also opportunities to learn and grow—both for us and our gardens.

As you plan your garden's layout and future rotations, consider each plant's needs and what it leaves behind in the soil. This thoughtful planning is akin to nurturing a community where every member contributes positively, ensuring long-term health and prosperity.

A Call to Action

Let this knowledge empower you. Start small if needed, perhaps by rotating just a few types of plants and observing the changes in their growth and health. Share your experiences with fellow gardeners, fostering a community spirit that thrives on shared knowledge and mutual support.

Your efforts to maintain soil health through rotation will ensure that your raised-bed garden continues to flourish. By embracing these principles, you honor the creation around you, promoting a cycle of life that nurtures and sustains itself year after year.

CHAPTER 13

Better Together: The Magic of Companion Planting

Mara stood in the midst of her garden, hands on her hips, surveying the small kingdom of green that sprawled before her. The sun was just beginning to dip below the horizon, casting long shadows across rows of tomatoes and beans. It was a quiet evening; the only sounds were the gentle rustling of leaves and the distant calls of birds returning to their nests.

She remembered her grandfather's garden—the way he used to hum old tunes as his large, weathered hands moved expertly through the soil. He had an uncanny understanding of which plants thrived together, a knowledge that seemed woven into his very being. Now Mara, with her own plot of land, hoped to mirror that harmony. But pests had become a relentless issue, nibbling away at her hard work.

As she walked between the rows, she brushed against a tomato plant; its sharp, distinctive smell rose in the air, mingling with the earthy scent of damp soil. Her thoughts drifted to something she had read about companion planting—how certain plant combinations could deter pests naturally. Could marigolds really keep away the aphids? Would planting basil alongside her tomatoes ward off other invaders?

She knelt beside a row where she had planted garlic some weeks back next to her roses on a whim after reading they could repel pests that favored roses. The garlic was thriving, robust and vibrant. It seemed less like coincidence now and more like quiet proof that there was truth in those old gardening tales.

The breeze picked up slightly, sending a chill down Mara's spine as she considered experimenting further with these natural partnerships. There was something deeply satisfying about solving problems with nature's own tools—about not relying on harsh chemicals but instead creating a balance that perhaps could heal more than just plant ailments.

A robin landed nearby, tilting its head curiously at Mara before hopping closer to peck around a squash plant. She watched it for a moment and thought about how companion planting might also support beneficial insects and birds—another layer of harmony in this complex tapestry.

As twilight deepened and stars began to prick at the darkening sky, Mara filled with resolve. Tomorrow would be for research and planning; for sketching out new configurations in her garden journal by lamplight. But tonight was for dreaming—for imagining her garden not just as it was but as it could be: vibrant with life both seen and unseen.

Could this natural approach really shift the balance in her favor? Could these old ways weave new strength into her modern-day garden?

Unearth the Secrets of a Thriving Garden

Gardening in urban spaces can often feel like a battle against limited space and challenging conditions. Yet, even in the most compact raised bed, a well-planned approach to planting can transform your garden into a robust, pest-resistant oasis. This magic is largely thanks to the ancient practice of companion planting, a method that not only enhances plant growth but also fosters a natural ecosystem right in your backyard.

At its core, **companion planting** involves strategically placing plants together that benefit each other in various ways. This could mean they deter pests naturally, attract beneficial insects, or improve soil quality through their unique root systems or nitrogen-fixing abilities. It's about creating plant partnerships that bring out the best in each other, thereby maximizing the productivity and harmony of your garden.

For those new to this concept, understanding which plants are compatible is crucial. Certain combinations can discourage harmful pests without the need for chemical insecticides. For example, marigolds emit a scent that repels nematodes and other pests while attracting pollinators like bees and butterflies. Similarly, planting basil near tomatoes not only improves their flavor but also repels flies and mosquitoes.

However, it's equally important to recognize what not to plant together. Some plants can have adverse effects on each other due to their nutritional demands or root structures. For instance, keeping onions away from beans is advisable as they can inhibit each other's growth.

Harness Nature's Little Helpers

By embracing companion planting, you also pave the way for *beneficial insects* to thrive. These insects are natural predators to common pests and play a crucial role in maintaining the health of your garden. Plants like lavender and dill attract ladybugs which feed on aphids, a common threat to many garden plants.

Creating an environment conducive to these helpful creatures involves more than just selecting the right plants. It's about designing your garden as a welcoming habitat for them—incorporating elements such as shallow water sources or specific flowering plants that provide nectar throughout the growing season.

Plan Your Plant Partnerships

With a clear understanding of beneficial plant relationships and insect attractions, planning your garden layout becomes an exciting puzzle to solve. This involves thinking about plant heights, sunlight needs, and root depths to ensure every plant has its place without overshadowing its neighbors.

Forming "guilds" or groups of complementary plants can simplify this process. For example, taller sunflowers can provide shade needed by lower-growing lettuce while their stalks serve as natural trellises for climbing beans.

Keep Your Garden in Harmony

Regular monitoring and maintenance play pivotal roles in sustaining the harmony within your companion-planted garden. Observing how plants interact throughout the seasons will help you make necessary adjustments—whether it's changing plant locations based on sun exposure or replacing certain combinations that don't work well.

Remember, every gardening season brings new lessons; what didn't thrive one year might flourish the next with some tweaks. This adaptive approach ensures your raised bed remains not only productive but also a dynamic part of your urban landscape.

Cultivate & Thrive: A Step-by-Step Guide to Companion Planting

The goal here is simple yet transformative: create a thriving garden ecosystem where plants act as mutual supporters in growth and pest management. Let's break down this process:

1. **Understand Plant Combinations**: Start by learning which plants make good neighbors and which should be kept apart. Integrate plants that naturally repel pests while enhancing each other's growth.
2. **Boost Beneficial Insects**: Incorporate plants known for attracting beneficial insects. Plan your garden layout to include these as integral parts of your ecosystem.
3. **Plan Harmonious Partnerships**: Strategically design your garden space considering light, space, and companion benefits to foster optimal growth conditions.
4. **Maintain Garden Harmony**: Keep an eye on your garden's dynamics; adjust as necessary to maintain balance between plant interactions ensuring productivity remains high.

Through this guide, transform your raised bed into a resilient mini-ecosystem where every plant plays its part beautifully—a testament to nature's inherent wisdom in fostering interdependence among living things.

Discovering Plant Synergies

Companion planting is not just a method; it's a partnership strategy in the garden. Think of it as finding the best roommate for your plants. Some plants, when grown together, can help each other thrive by repelling pests naturally. This method reduces the need for chemical pesticides, making your garden more sustainable and eco-friendly.

Tomatoes and basil are classic companions. The strong scent of basil repels harmful insects like thrips and mosquitoes while attracting beneficial pollinators that boost tomato yields. This pairing not only maximizes space but also enhances the flavor of tomatoes, making this duo a favorite among culinary gardeners.

In another example, marigolds are not just pretty flowers; they are warriors in disguise. Planting marigolds among vegetables helps repel nematodes and other pests that could otherwise harm your crops. Their bright blooms act as a beacon for good bugs while deterring the harmful ones.

Imagine your garden as a community where every plant plays a role in supporting its neighbor. Carrots and onions are good neighbors because onions release substances that deter carrot flies. This mutual support system reduces stress on each plant, allowing them to grow stronger and healthier.

Key point: Learning which plant combinations benefit growth and repel pests can transform your garden into a flourishing ecosystem with minimal pest issues.

Boosting Beneficial Insect Populations

Companion planting does more than just repel pests; it can also create an environment that attracts beneficial insects. These insects are nature's pest control agents, helping to maintain the balance in your garden without chemicals.

For instance, planting flowers like cosmos or yarrow near vegetable beds can attract ladybugs and lacewings, which feed on aphids—a common pest on many vegetables and flowers. These predatory insects reduce aphid populations naturally, promoting healthier plant growth.

Herbs such as dill and fennel attract wasps that are harmless to humans but deadly to many garden pests. These wasps lay their eggs inside caterpillars and other insect pests, controlling their populations effectively. Incorporating these herbs into your garden not only adds variety but also fortifies your natural defense system against pests.

Consider how a single flower in your garden can serve as a powerful magnet for pollinators like bees and butterflies while simultaneously warding off invaders. This dual role makes certain plants indispensable in the ecological tapestry of your garden.

Can you imagine how incorporating these practices might change not only the health of your plants but also their productivity?

Harmonizing Your Garden Through Strategic Plant Partnerships

The Companion Planting Process Model (CPPM)

This model provides a framework for selecting companion plants to maximize garden health and productivity based on specific criteria such as nutrient needs, pest control, and physical characteristics.

Decision Trees for Companion Selection

At each decision point in our model, consider factors such as sun exposure, watering needs, and growth periods. For example, if you're planning to grow leafy greens which require ample shade, pairing them with taller plants like sunflowers or corn can provide the necessary cover, reducing water evaporation from the soil.

Complementary Nutrient Uptake

Plants like beans fix nitrogen in the soil, which benefits nitrogen-loving plants like corn. This relationship not only improves soil fertility but also reduces the need for chemical fertilizers. Each plant extracts different nutrients at varying depths, which prevents depletion of any single nutrient.

Pest Control Synergies

The model emphasizes plant relationships that naturally repel pests or attract their predators. For instance, garlic planted near roses deters aphids; similarly, chives improve the taste and growth of carrots while keeping pests at bay.

Spatial Efficiency

Utilize vertical space by growing climbers like cucumbers near sturdy plants like corn which act as natural trellises. This efficient use of space is particularly useful in small gardens or urban settings where ground space is limited.

Dynamics Over Time

As plants grow according to this model, their interactions can lead to increased yield and decreased pest problems over time. Feedback loops form where healthier plants bolster their neighbors' defenses against pests through stronger scents or more vigorous growth.

Exploring these relationships through companion planting not only makes your gardening practice more sustainable but also enriches the biodiversity of your urban garden setting

As we reflect on the enriching practice of companion planting, it becomes clear that understanding the relationships between different plants is not just about improving yields; it's about fostering a thriving ecosystem right in our urban gardens. **Plant partnerships** are fundamental to enhancing the health and productivity of your garden. By choosing the right combinations, you not only discourage pests naturally but also promote a more robust population of beneficial insects.

Mutual benefits arise when plants that complement each other are grown together. For instance, some plants repel pests that might otherwise devastate their neighbors, while others attract beneficial insects that aid in pollination or pest control. This natural method of maintaining plant health reduces the need for chemical pesticides, which can be harmful to both the environment and human health.

Moreover, the practice of companion planting goes beyond pest management and pollination. It includes improving soil quality through natural processes like nitrogen fixation, which is essential for building fertile, healthy soil capable of supporting diverse plant life. Strategic plant placement can also optimize space and sunlight exposure, ensuring that each plant receives the care it needs without hindrance.

Embracing these principles not only enhances your garden's productivity but also contributes to a healthier environment. As urban gardeners, we have a unique opportunity to influence our immediate ecological landscape positively. Implementing companion planting is a step towards more sustainable gardening practices that respect and enhance natural processes.

Let us be encouraged by the potential of companion planting to transform our gardening practices. By adopting these strategies, we not only improve our own gardens but also contribute to a greener, more sustainable urban community. Let's continue to learn, experiment, and grow together in harmony with nature's designed patterns.

CHAPTER 14

From Novice to Green Thumb: Your Sustainable Gardening Journey

In the quiet suburb where small gardens flourished like patches of hope, Sarah stood alone in her backyard, hands on her hips, surveying the compact space that she had earmarked for a transformation. The sun was high and cast a harsh light over her shoulder, highlighting the neglected patches of earth that awaited her attention. It was a Saturday morning, and the neighborhood thrummed with the sound of lawn mowers and children's laughter.

As she pondered where to begin, Sarah's mind wandered to her grandmother's garden; a lush, vibrant plot teeming with tomatoes and cucumbers that seemed to burst forth from the soil in defiance of the cramped space they were allotted. She remembered how her grandmother spoke to the plants, believing they could hear her encouragement or sense her frustration. Now faced with her own plot of land, however small and urban-bound it might be, Sarah felt a twinge of doubt—could she replicate that lush garden within these modern confines?

The clatter of a shovel falling from the fence snapped Sarah back to reality. She picked it up, feeling its solid handle grounding her resolve. Turning over the first clump of earth felt like both an intrusion and a beginning. As she worked, dirt found its way under her nails and sweat formed beads along her hairline.

Her neighbor Tom leaned over the fence, his eyes squinting against sunlight. "Starting your garden project?" he asked.

"Yes," Sarah replied with a forced smile. "I'm trying to go sustainable—organic practices only."

Tom nodded approvingly before offering some advice about composting—a practice he'd mastered over years tending his own garden plot next door.

As she returned to digging, thoughts about sustainability circled in Sarah's mind like birds overhead. It wasn't just about avoiding chemicals; it was about nurturing something pure from the ground up. Each shovel-full of earth turned was an act toward healthier living—not just for herself but for anyone who would share in this garden's bounty.

Later that afternoon as shadows grew long and muscles sore, Sarah paused again. She wiped sweat from her brow with a dirty forearm and surveyed what little progress seemed visible after hours of labor. It struck her then how gardening was less about immediate results and more about ongoing effort—the patience required not just to plant but also to cultivate growth day by day.

Would this patch ever resemble those vibrant images captured in gardening books or would nature have its own ideas? How much could one small space teach about life itself?

Unlocking the Secrets of Your Garden's Potential

Embarking on the journey of creating a sustainable garden in an urban setting can be transformative. This venture not only cultivates fresh produce but also nurtures the soul. As we delve into the essentials of gardening, remember that each step you take enriches your connection with nature and fortifies your self-sufficiency.

Equipping yourself with essential knowledge and tools is the cornerstone of any successful gardening endeavor. Understanding soil composition, water requirements, and plant compatibility are fundamental to fostering a thriving garden. Tools like trowels, pruning shears, and watering cans are not just equipment; they are extensions of your intent to nurture life from the earth. Embrace these tools as partners in your journey.

The empowerment derived from **sustainable and organic gardening practices** cannot be overstated. By choosing to avoid synthetic chemicals and pesticides, you are making a conscious decision to protect the environment and your health. This approach aligns with a stewardship mindset, echoing the biblical call to "tend and keep" the earth (Genesis 2:15). It's a practice that not only benefits your garden but also contributes to a healthier planet.

A profound aspect of gardening is its ability to **foster a deeper appreciation for the natural world**. The process of planting seeds, nurturing them, and watching them grow is a vivid illustration of life's cycles. It reminds us of our place within these cycles and our responsibility towards them. In cultivating a garden, you connect with the rhythm of life, experiencing firsthand the miracles that unfold daily under the sun.

Throughout this book, we've explored various aspects of urban gardening—from selecting the right soil to managing small spaces effectively. These lessons converge here, empowering you to apply these insights practically. Whether you're arranging your first raised bed or fine-tuning the pH levels of your soil, each action is a step toward sustainability.

Reflecting on what we've learned can reinforce our commitment to these practices. Gardening is not just about harvesting food; it's about sowing seeds of change—both in our gardens and in our communities. By sharing your harvest or gardening knowledge with neighbors, you sow seeds of community and resilience.

As we near the end of this guide, remember that every plant in your garden started as a seed—small yet brimming with potential. Similarly, your journey as a gardener began with a single step: deciding to start. With each new season, refine your techniques, expand your knowledge base, and continue to foster love for the soil beneath your feet and the food it brings forth.

Through this process, you do more than grow vegetables; you cultivate life itself.

Equip Yourself with the Knowledge and Tools to Start Your Gardening Adventure

Starting a garden is like opening a book for the first time — each page a step toward a new world of knowledge. The initial step for any aspiring gardener is to arm oneself with the essential tools and knowledge. Understanding soil types, essential nutrients, and the right tools can significantly ease your gardening initiation.

Tools are your best allies in this venture. A sturdy spade, a reliable rake, and a set of gardening gloves are just the beginning. These tools do not just serve their physical purpose but also symbolize your readiness to dig deep and cultivate not only plants but also your capabilities as a gardener.

The type of soil in your garden dictates the success of your plants. Sandy soil drains quickly but doesn't hold nutrients well, while clay soil holds nutrients but may not drain well. Knowing this helps in choosing the right plants or improving the soil structure with organic matter like compost. This addition not only enriches the soil but also promotes sustainable practices by recycling kitchen waste.

Imagine if every seed you plant is a piece of knowledge you sow, growing into lush insights as you nurture it with attention and care. Gardening requires patience and persistence, attributes that also allow you to grow personally alongside your garden.

Equip yourself with basic gardening tools and an understanding of your environment to start your gardening journey successfully.

Find Empowerment and Inspiration in Sustainable and Organic Gardening Practices

Organic gardening transcends the act of planting; it's a commitment to sustaining life healthily and naturally. By choosing organic methods, you are deciding to protect the environment from harsh chemicals, thus ensuring that the food on your table is as wholesome as it gets.

The benefits of organic gardening are profound. Chemical-free soils produce healthier plants which in turn nourish healthier humans and animals. Furthermore, by avoiding synthetic chemicals, you reduce groundwater pollution, creating a healthier ecosystem for local wildlife.

One might wonder how personal empowerment fits into gardening. Every seed planted organically is an assertion of control over one's health and environment. This act empowers individuals by connecting them deeply with nature's cycles and teaching resilience through seasonal changes.

Think of organic gardening as crafting a legacy — not just for oneself but for future generations who will benefit from sustainable practices established today.

How can embracing organic principles transform not just your garden, but also your life?

Foster a Deeper Appreciation for Growing Food and Connecting with Nature Through Gardening

Gardening is more than just an activity; it's an interaction with nature that teaches us about growth, renewal, and resilience. Each plant tells a story through its lifecycle — from sprouting seedlings to flourishing greenery offering fruits or flowers.

Understanding plant life cycles connects us intimately with nature's rhythms. We learn patience as we watch tomatoes ripen or basil leaves multiply. This connection deepens our respect for nature's timings, distinct from our fast-paced modern lives.

Moreover, harvesting food that one has grown can transform eating into an experience of gratitude and pride. There's something profoundly satisfying about cooking a meal seasoned with herbs picked from one's garden; it's a tangible result of care and effort.

Imagine nurturing not just plants but relationships through shared garden tasks or meals prepared from harvests. These experiences can strengthen community bonds and foster mutual support among friends and family members.

By equipping ourselves with gardening knowledge, embracing sustainable practices, and connecting deeply with nature, we cultivate more than food — we grow life itself.

As we reach the end of our journey together, it's important to reflect on the strides you've made from a novice to someone with a budding green thumb. The path to establishing your raised-bed garden has been filled with learning and growth, not just for your plants but for you as an individual. Embracing the challenge of starting with limited space and knowledge, you have now cultivated a vibrant space of sustainability and health.

Equipping yourself with the right tools and knowledge was your first step toward success. Remember, every seed planted and every soil turned was a step closer to understanding the rhythms of nature. This process wasn't just about growing plants but also about nurturing your capability to care for creation, resonating deeply with the stewardship we are called to uphold.

Finding **empowerment in sustainable practices** has hopefully shifted your perspective on what it means to interact with the earth. Organic gardening is more than a method—it's a testament to the respect and care for the natural world that is advocated throughout scripture. As you continue to tend your garden, let each act of watering, pruning, and harvesting be a reflection of love and respect for creation.

The joy of harvesting your own produce is unmatched, fostering a **deeper appreciation for nature** and its Creator. These moments of harvest are not just about enjoying fresh, healthy food but also celebrating the miraculous processes set forth by God. Let these experiences remind you of the parables of growth in faith; just as your plants need water and light, your spirit thrives on nourishment from the word of God.

This book aimed to demystify gardening in small spaces while encouraging you to embark on this green journey. Hopefully, it has served as a beacon of hope and guidance as you transformed bare patches into lush gardens brimming with life. Continue to share this bounty—both the literal fruits and vegetables and the spiritual and communal wealth it brings.

As you move forward, remember that each season brings new lessons and opportunities for growth—both in your garden and in your life. May your hands always be dirty from the soil, your heart full from the connection to Earth, and your spirit enriched by the continuous cycle of growth and renewal.

Epilogue

Cultivating Growth: From Seeds to Harvest

As we draw this guide to a close, let us reflect on the journey we have embarked upon together. Throughout this book, we've explored not only the practical steps necessary to create and nurture a raised-bed garden in urban spaces but also the deeper implications of gardening—how it nurtures not just our bodies with fresh produce but our spirits and communities as well.

The practical applications of what you've learned are vast. You can now transform even the smallest plot or balcony into a thriving garden that provides not only sustenance but also a sense of accomplishment and peace. This is your opportunity to apply these principles in your daily life, creating a space that reflects both care for the environment and for your own well-being.

We've covered essential topics such as selecting the right location for your garden beds, choosing soil and compost, deciding on plants that thrive in your specific conditions, managing pests naturally, and harnessing organic methods to boost your garden's health and output. These elements are crucial in ensuring that your gardening efforts are fruitful.

To **implement what you've learned**, start small. Choose one or two key takeaways from each chapter and apply them. Whether it's testing your soil before planting, setting up a proper drainage system in your raised beds, or experimenting with companion planting to naturally deter pests—each action you take is a step towards becoming a more proficient gardener.

It's important to acknowledge that while this guide provides a comprehensive introduction to raised-bed gardening, there is always more to learn. The world of gardening is as vast as nature itself, and there are always new techniques, tools, and plant varieties to explore. I encourage you to continue educating yourself through other resources—books, community classes, or online forums—and remain open to new ideas and innovations that can enhance your gardening experience.

I urge you **to take action** today. Don't let what you've learned sit idle. Plant those seeds, both literal and metaphorical, and watch as they grow into something beautiful and nourishing under your care.

May this book serve not only as a guide but as an inspiration for you to embrace gardening as a part of your lifestyle. Remember that every small patch of soil holds the potential for growth not just of plants but of the soul. Let your garden be a testament to what can be achieved with faith, dedication, and love.

As we part ways in this written form, carry forward the message that even in the smallest spaces, great things can grow. May you find joy in every sprout and strength in every challenge.

"The glory of gardening: hands in the dirt, head in the sun, heart with nature. To nurture a garden is to feed not just on the body, but the soul." — Alfred Austin

May your gardens flourish and your hearts be full as you walk this path of growth and discovery.

"If you enjoyed this book, would you be so kind as to help us remain in the top listing by leaving a positive Book review."